THE AFS GUIDE TO FISHERIES EMPLOYMENT
SECOND EDITION

D0809634

THE AFS GUIDE TO FISHERIES EMPLOYMENT

is a special project of and was partly funded by the

Education Section and Student Subsection of the American Fisheries Society

THE AFS GUIDE TO FISHERIES EMPLOYMENT
SECOND EDITION

Edited by

David A. Hewitt
Department of Fisheries Science
Virginia Institute of Marine Science
College of William and Mary

William E. Pine, III
Department of Fisheries and Aquatic Sciences
University of Florida

Alexander V. Zale
U.S. Geological Survey
Montana Cooperative Fishery Research Unit
Department of Ecology
Montana State University

American Fisheries Society
Bethesda, Maryland

Suggested citation formats are

Entire book

Hewitt, D. A., W. E. Pine, III, and A. V. Zale, editors. 2006. The AFS guide to fisheries employment, 2nd edition. American Fisheries Society, Bethesda, Maryland.

Chapter in book

Willis, D. W., and C. B. Grimes. 2006. Developing your knowledge, skills, and professionalism as an undergraduate. Pages 1–12 *in* D. A. Hewitt, W. E. Pine, III, and A. V. Zale, editors. 2006. The AFS guide to fisheries employment, 2nd edition. American Fisheries Society, Bethesda, Maryland.

Printed in the United States of America on acid-free paper.

Contribution 2741 of the Virginia Institute of Marine Science, College of William and Mary

Library of Congress Control Number 2006932274

ISBN 1-888569-86-7

American Fisheries Society web site address: *www.fisheries.org*

American Fisheries Society
5410 Grosvenor Lane, Suite 110
Bethesda, Maryland 20814-2199
USA

CONTENTS

Contributors..vii

Preface..xi

Chapter 1. Developing Your Knowledge, Skills, and Profession-
alism as an Undergraduate...1
David W. Willis and Churchill B. Grimes

Chapter 2. Developing an Effective Resume or Curriculum
Vitae...13
Trent M. Sutton, Rebecca A. Zeiber, and Emmanuel A. Frimpong

Chapter 3. Pursuing Graduate Studies in Fisheries.......................39
Alexander V. Zale

Chapter 4. Fisheries Employment in State Agencies......................57
Richard T. Eades

Chapter 5. Fisheries Employment in the U.S. Federal Govern-
ment..79
Robert L. Simmonds, Jr. and Mary C. Fabrizio

Chapter 6. Academic Positions in Fisheries Science......................97
Brian R. Murphy

Chapter 7. Fisheries Employment in Cooperative Research
Units: Where Agency Meets Academia...109
Thomas J. Kwak and F. Joseph Margraf

Chapter 8. Employment in Aquaculture..123
Anita M. Kelly

Chapter 9. Fisheries Employment in Canada................................135
Steven J. Cooke and Scott G. Hinch

Chapter 10. International Fisheries Employment...........................151
Richard A. Neal

Chapter 11. Fisheries Employment with Nongovernmental Organizations..165
William E. Pine, III and Kenneth M. Leber

Chapter 12. Private Consulting in Fisheries Science.............................175
Donald D. MacDonald, Forrest Olson, and Andrew J. Loftus

Chapter 13. Advancing to a Career in Fisheries Administration........183
Steve L. McMullin and Christopher Hunter

Chapter 14. Equal Opportunities in Fisheries Employment.................205
Columbus H. Brown and Essie C. Duffie

Chapter 15. The American Fisheries Society: A Framework for Professionalism..215
Ghassan (Gus) N. Rassam

CONTRIBUTORS

Columbus H. Brown (Chapter 14): U.S. Fish and Wildlife Service, 1875 Century Boulevard, Suite 205, Atlanta, Georgia 30345, USA.

Steven J. Cooke (Chapter 9): Department of Biology and Institute of Environmental Science, Carleton University, 1125 Colonel By Drive, Ottawa, Ontario K1S 5B6, Canada.

Essie C. Duffie (Chapter 14): National Oceanic and Atmospheric Administration, National Marine Fisheries Service, Southeast Fisheries Science Center, 75 Virginia Beach Drive, Miami, Florida 33149, USA.

Richard T. Eades (Chapter 4): Nebraska Game and Parks Commission, P.O. Box 30370, Lincoln, Nebraska 68503, USA.

Mary C. Fabrizio (Chapter 5): Department of Fisheries Science, Virginia Institute of Marine Science, College of William and Mary, P.O. Box 1346, Gloucester Point, Virginia 23062, USA.

Emmanuel A. Frimpong (Chapter 2): Department of Forestry and Natural Resources, Purdue University, 195 Marsteller Street, West Lafayette, Indiana 47907, USA.

Churchill B. Grimes (Chapter 1): National Oceanic and Atmospheric Administration, National Marine Fisheries Service, Southwest Fisheries Science Center, 110 Shaffer Road, Santa Cruz, California 95060, USA.

David A. Hewitt (Coeditor): Department of Fisheries Science, Virginia Institute of Marine Science, College of William and Mary, P.O. Box 1346, Gloucester Point, Virginia 23062, USA.

Scott G. Hinch (Chapter 9): Department of Forest Sciences, University of British Columbia, 2424 Main Mall, Vancouver, British Columbia V6T 1Z4, Canada.

Christopher Hunter (Chapter 13): Montana Department of Fish, Wildlife and Parks, P.O. Box 200701, Helena, Montana 59620, USA.

Anita M. Kelly (Chapter 8): Fisheries and Illinois Aquaculture Center, Southern Illinois University, Mailstop 6511, Carbondale, Illinois 62901, USA.

Thomas J. Kwak (Chapter 7): U.S. Geological Survey, North Carolina Cooperative Fish and Wildlife Research Unit, North Carolina State University, Campus Box 7617, Raleigh, North Carolina 27695, USA.

Kenneth M. Leber (Chapter 11): Mote Marine Laboratory, Center for Fisheries Enhancement, 1600 Ken Thompson Parkway, Sarasota, Florida 34236, USA.

Andrew J. Loftus (Chapter 12): Andrew Loftus Consulting, 3116 Munz Drive, Suite A, Annapolis, Maryland 21403, USA.

Donald D. MacDonald (Chapter 12): MacDonald Environmental Sciences, Ltd. and Sustainable Fisheries Foundation, #24-4800 Island Highway North, Nanaimo, British Columbia V9T 1W6, Canada.

F. Joseph Margraf (Chapter 7): U.S. Geological Survey, Alaska Cooperative Fish and Wildlife Research Unit, University of Alaska Fairbanks, P.O. Box 757020, Fairbanks, Alaska 99775, USA.

Steve L. McMullin (Chapter 13): Department of Fisheries and Wildlife Sciences, Virginia Polytechnic Institute and State University, 100 Cheatham Hall, Mail Code 0321, Blacksburg, Virginia 24061, USA.

Brian R. Murphy (Chapter 6): Department of Fisheries and Wildlife Sciences, Virginia Polytechnic Institute and State University, 100 Cheatham Hall, Mail Code 0321, Blacksburg, Virginia 24061, USA.

Richard A. Neal (Chapter 10): National Oceanic and Atmospheric Administration, National Marine Fisheries Service, Southwest Fisheries Science Center, 8604 La Jolla Shores Drive, La Jolla, California 92037, USA.

Forrest Olson (Chapter 12): CH2M Hill, Inc., P.O. Box 91500, Bellevue, Washington 98009, USA.

William E. Pine, III (Coeditor and Chapter 11): Mote Marine Laboratory, Center for Fisheries Enhancement, 1600 Ken Thompson Parkway, Sarasota, Florida 34236, USA. (Present address: Department of Fisheries and Aquatic Sciences, University of Florida, 7922 NW 71st Street, Gainesville, Florida 32653, USA.)

Ghassan (Gus) N. Rassam (Chapter 15): Executive Director, American Fisheries Society, 5410 Grosvenor Lane, Suite 110, Bethesda, Maryland 20814, USA.

Robert L. Simmonds, Jr. (Chapter 5): U.S. Fish and Wildlife Service, Carterville Fishery Resources Office, 9053 Route 148, Suite A, Marion, Illinois 62959, USA.

Trent M. Sutton (Chapter 2): Department of Forestry and Natural Resources, Purdue University, 195 Marsteller Street, West Lafayette, Indiana 47907, USA.

David W. Willis (Chapter 1): Department of Wildlife and Fisheries Sciences, South Dakota State University, P.O. Box 2140B, Brookings, South Dakota 57007, USA.

Alexander V. Zale (Coeditor and Chapter 3): U.S. Geological Survey, Montana Cooperative Fishery Research Unit, Department of Ecology, Montana State University, Bozeman, Montana 59717, USA.

Rebecca A. Zeiber (Chapter 2): Department of Forestry and Natural Resources, Purdue University, 195 Marsteller Street, West Lafayette, Indiana 47907, USA.

PREFACE

The first edition of *An AFS Guide to Fisheries Employment* was edited by Tracy Hill and Robert Neumann and published in January 1996 as a special project of the Student Subsection of the Education Section of the American Fisheries Society (AFS). As the tenth anniversary of the *Guide*'s publication approached, the Education Section and the Student Subsection decided that it was time for an update and expansion of the earlier edition. The first edition of the *Guide* was a valuable resource for students and young professionals pursuing careers in fisheries science and related disciplines. We hope that the second edition will continue to serve these individuals—those for whom the task of laying out a road map to satisfying and rewarding employment is often most challenging—but we also hope that it will be useful for a wider audience. We have sought to include chapters in the second edition that cover all stages of common careers in fisheries, from undergraduate development to advancement into administrative positions. In addition, the chapters include more information about fisheries employment in marine and coastal environments.

As in the first edition, the bulk of the chapters are dedicated to providing practical advice on important standard topics, such as getting started with a solid undergraduate education (Chapter 1), designing an effective resume or curriculum vitae (Chapter 2), getting into graduate school (Chapter 3), and finding and landing a job with a state or federal agency (Chapters 4 and 5). Chapters 6 and 7 discuss employment in an academic setting, whether it be a typical college or university faculty appointment or a position in a Cooperative Research Unit. Chapter 8 describes what to expect in, and the education needed for, a variety of job opportunities in the aquaculture field. Other chapters cover employment outside of the United States (Chapters 9 and 10), opportunities available with nongovernmental organizations (Chapter 11) and private consulting firms (Chapter 12), and things to know and consider about moving into administrative positions (Chapter 13). The final two chapters discuss the rights and responsibilities we all have concerning equal opportunity employment (Chapter 14) and how

AFS can help individuals become consummate fisheries professionals (Chapter 15).

The chapters in the second edition of the *Guide* all speak to a common theme—that fisheries science is a diverse, interdisciplinary, and rapidly changing field. Alphonse Karr wrote in 1849, "The more things change, the more they stay the same." Karr presumably had no notion of the pace of change that the world is experiencing now, and while his words are still true in many cases, the accuracy of the phrase is being tested. Similar to technology, and in many ways because of it, the science and management of fisheries and aquatic resources are changing in ways that most of the retiring scientists of our day never envisioned. Indeed, we are being swept along with the current.

Coincident with changes in the science of the profession, the landscape of fisheries employment is changing as well. For example, as Steve McMullin and Chris Hunter note in Chapter 13, the Baby Boomers are on their way out and a new generation of fisheries scientists will be stepping up to take the lead in the 21st century. Some things related to employment in fisheries will probably never change, such as the tremendous value of networking in a field that always seems to remain small. However, given the changes that are occurring and will continue to occur, a commitment to life-long learning and an ability to adapt to new technologies and research directions will be essential qualities of the fisheries professional of the future.

Many of the chapters in the second edition are based on chapters in the first edition that were written by other authors. Accordingly, the authors of the second edition chapters and the editors acknowledge the contributions of those professionals: Bruce W. Menzel and Stephen A. Flickinger (Chapter 1); Richard L. Noble and Michael J. Van Den Avyle (Chapter 2); Donald J. Orth and Ira R. Adelman (Chapter 3); Linda Erickson-Eastwood and Fred A. Harris (Chapter 4); Neil B. Armantrout and Hannibal Bolton (Chapter 5); Nick C. Parker and Arden J. Trandahl (Chapter 8); Bruce A. Barton (Chapter 9); Gene R. Huntsman (Chapter 10); Andrew J. Loftus (Chapter 11); Christine M. Moffitt and Cay C. Goude (Chapter 14); and Paul Brouha and Robert L. Kendall (Chapter 15).

In addition to the contributions of the authors, the publication of this second edition of the *Guide* reflects the hard work and support of numerous members of the Education Section and the Student Subsection and other members of AFS. We are especially indebted to the individuals who graciously provided thorough reviews that significantly strengthened the chapters. For their volunteer efforts as peer reviewers we thank Ira R. Adelman, Jim Colvocoresses, Steven J. Cooke, John Copeland, Louis Daniel, Carol Endicott, Lynn Fegley, William G. Franzin, Timothy B. Grabowski, Amy L. Harig, Joe Hightower, Michael Hirshfield, Wayne A. Hubert, Don Jackson, Cecil A. Jennings, Sharon Kramer, Bill Manci, Steve L. McMullin, Gil McRae, Dirk Miller, Thomas J. Miller, Edward O. Murdy, Kevin L. Pope, James B. Reynolds, Frederick S. Scharf, Mary Schilling, Dana Schmidt, Carl B. Schreck, Paul Seidel, Roy A. Stein, Thomas N. Todd, Mike Van Den Avyle, Doug Vaughan, Craig A. Watson, and Gwen White.

David A. Hewitt
William E. Pine, III
Alexander V. Zale

Chapter 1

Developing Your Knowledge, Skills, and Professionalism as an Undergraduate

DAVID W. WILLIS AND
CHURCHILL B. GRIMES

The life of a fisheries biologist: being outdoors with the sun in your face and the wind in your hair, working with fish, not a care in the world? Well, that is only partially correct. Fisheries professionals often are just as involved with people and people management as they are with the fish themselves. Of course, the extent to which a fisheries professional works with the public will vary depending on his or her position, such as whether it empha-sizes research, management, or policy. Students often are sur-prised to learn that undergraduate fisheries education typically includes substantial credits in mathematics and the physical and social sciences, in addition to the expected biological sciences. People who like to work outdoors but have little aptitude in sci-ence may be wise to choose another profession, and let their out-door activities be an avocation rather than a vocation.

To help understand the need for a broad, science-based edu-cation, consider the following description. A *fishery* consists of three interacting components (Figure 1.1). First, fishes are part of the *biota* (living organisms). The biota includes both target spe-cies, such as sport fish or commercially harvested fish or inverte-brates (e.g., lobsters and crabs), and non-target species, such as prey fish, aquatic insects, or bacteria. *Habitat*, the second part of the fishery triad, is a necessary requirement for all living organ-isms and includes both the living and nonliving portions of the

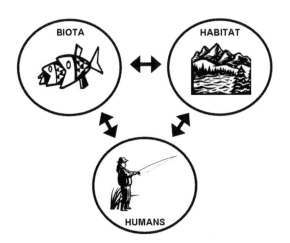

Figure 1.1. A *fishery* is composed of three interacting elements: habitat, biota (organisms), and humans. Undergraduate education needs to encompass all three elements. Some classes may be quite specific to one element, such as a fish taxonomy class (biota) or an economics class (humans). Other classes will cover all three elements, such as a fisheries management class.

environment. *Humans* are the third component of a fishery. Humans are both users of fishes themselves (sport or commercial fishing) and competitors for water. Many human activities, such as manufacturing or agriculture, require substantial amounts of water and may degrade or destroy critical habitats. *Fisheries management* is the manipulation of the three elements in a fishery to meet intended and desirable objectives.

Thus, a fisheries education must prepare students to understand not only the fishes themselves, but also their habitat requirements and the human dimensions of fisheries. In addition to math and science coursework, communications skills and knowledge of social sciences and humanities are essential for success in fishery management because of the political landscape upon which fisheries research and management are superimposed. Attitudes toward natural resource use and science have begun to change in recent years. For example, there is less concentration on the population ecology of single species and increasing emphasis on ecosystem-based management. Similarly, commercial exploitation is challenged by changing societal values and emphasis on non-consumptive uses and conservation (including non-game and threatened or endangered species). As a result, there is a broad diversity in employment opportunities in marine and freshwater fisheries.

The purpose of this chapter is to provide information that will allow students to better prepare themselves for careers in fisheries. We discuss the knowledge base that should be obtained through formal coursework, provide information that will allow students to gain the practical experience needed for these disciplines, and then emphasize the role of professionalism in both education and eventual employment. We note that the educational paths for students seeking eventual employment in freshwater fisheries and marine fisheries will often differ substantially, but the end results are similar.

Career Opportunities

Students can more easily choose an appropriate educational path if they fully understand the career options available to them and follow their personal interests. This guide covers the primary career paths followed by freshwater and marine fisheries professionals and emphasizes potential employers. Each career path will result in unique educational requirements, with graduate school often being part of the overall education (see Chapter 3).

At the undergraduate level, students likely will need to make a choice between a more general biological undergraduate degree (which may be labeled environmental science, natural resources management, ecology, conservation biology, etc.) and a more focused undergraduate education at a school having a traditional fisheries science program. Such traditional programs are common in many inland states and a few coastal states. Few focused undergraduate programs exist for marine fisheries. Most graduate programs in marine fisheries select students with general biological or mathematical backgrounds that have an interest in marine ecology. Inland fisheries graduate programs accept students from both traditional fisheries programs and general biological programs.

Educational Requirements

A traditional undergraduate degree in fisheries is one path by which to begin a career in fisheries. A biological degree, especially with a few of the right electives, may be just as good, especially if a student plans to subsequently attend graduate school. A fisheries degree will be useful for those students who plan to seek traditional fisheries employment (e.g., biologist or technician) after completion of only a Bachelor's degree.

The American Fisheries Society (AFS) maintains a program to certify Fisheries Professionals (see also Chapter 15). The list of required coursework for this certification program provides a guide to necessary coursework that students should consider as they select their undergraduate and graduate programs. While professional certification is not required for most employment opportunities, a few state agencies do require certification, others provide pay incentives for employees who are certified, and at least one state agency uses the minimum requirements of the AFS certification program as the minimum requirements to be considered for agency employment. Thus, we recommend that undergraduate students either complete the requirements or plan ahead to ensure that they can finish the requirements during their subsequent graduate education.

Students should complete a number of courses focused specifically on fisheries and aquatic sciences, at least one of which should cover principles of fisheries management. Students should also take courses that will provide a solid foundation in other biological sciences, such as evolutionary biology or ecology, and physical sciences, such as chemistry or geology. Importantly, students should select courses that are of particular interest to them whenever possible. Fisheries science and management are becoming more and more quantitative, so students should include sufficient coursework in mathematics and statistics. Communications skills are essential for fisheries professionals, so coursework in both writing and public speaking are encouraged. Finally, courses related to the human dimensions of fisheries should be included in undergraduate training. Human dimensions courses cover a wide array of topics, such as natural resource policy, law, ethics, public relations, and economics.

Based on the fishery triad (biota, habitat, and the human dimension), an undergraduate curriculum must lead to an understanding of habitat management, the human dimensions of fisheries management, and the fishes and other living organisms in the system. Moreover, the goal of undergraduate education at universities is to provide a broad education, including humanities and social sciences. No university is simply in the business of providing technical training in a narrow topic area. This philosophy of broad education is essential in fisheries and aquatic sciences.

Students should select an undergraduate program based on its ability to provide education in the diverse areas of responsibility typical of entry-level fisheries professionals. Graduates should strive to have an understanding of the following basic topics:

• Physical, chemical, and biological processes (including population, community, and ecosystem processes) in aquatic environments;

• The scientific method, experimental design, and sampling procedures;

• Library research and information retrieval;

• Water quality analysis and interpretation;

• Identification of common fishes and aquatic invertebrates, algae, and vascular plants;

• Fish biology (especially physiology, ecology, life history, and population dynamics);

• Fish anatomy and morphology;

• Fish sampling methods;

• Analysis of fish samples for length, weight, age, and sex;

• Estimation of fish population demographics and assessment of different levels and strategies of fishing on populations and prediction of future yields;

• Animal welfare concerns;

• Concepts of natural resource management and related social and economic issues;

• Basic techniques for management, summarization, and statistical analysis of data;

• Oral and written technical communication, popular writing, public speaking, and public relations (including conflict resolution);

• Fundamentals of budget preparation and management and employee supervision.

Not all undergraduate programs can provide background in all of these areas; however, the combination of education and experience gained by graduates should include most of these qualifications. Once again, however, students planning to attend graduate school will have additional years of coursework and experience to help acquire this knowledge base.

Useful Minors and Elective Credits

Most undergraduate programs have a degree of flexibility, so that at least a few of the required credits for graduation can be elective courses selected by the student. In addition, a minor in an area of interest may help expand your professional career opportunities, although the credits required for a minor may cause the student to exceed the minimum requirements for a Bachelor's degree. In such a situation, students will need to balance the costs and benefits of a potential additional semester against those of spending that time in a job or graduate program.

We recommend that students pursue minors or elective credits in their own areas of interest. If a student likes working with computers, a minor or at least some coursework in geographic information systems (GIS) would be quite valuable in obtaining future employment. Fisheries professionals use GIS in many aspects of their careers. For a student interested in natural resources law enforcement, coursework or a minor in criminal justice may be an option. If a student is interested in a career in fisheries research, bolstering his or her quantitative skills through additional coursework in mathematics and statistics would be prudent. For someone primarily interested in the human dimensions of fisheries, elective credits could be used to take additional courses in the social and political sciences. If a student plans to pursue a career in the private aquaculture or pond management industries, his or her biological knowledge should be complemented with coursework or a minor in business. Students interested in environmental consulting might consider courses in hydrology, water quality, and environmental law. If a student hopes to eventually have a career in fisheries administration, coursework in personnel management and administration will be quite helpful.

Students should keep in mind that modern fisheries science incorporates disciplines that have long been considered "basic" science, such as evolutionary or developmental biology, molecular biology/ecology, and behavioral science. Students with particular interests in these fields will be able to find opportunities to apply such interests and expertise toward fisheries-related issues. Individuals will be most successful in both their education and future career endeavors when they follow their personal interests.

Learning through Experience

Although some undergraduate coursework will include hands-on experience, most students that obtain fisheries employment will have acquired additional experience. The American Fisheries Society sponsors a program through which even high school students can gain fisheries field experience (Box 1.1). Undergraduate students can gain experience over and above that obtained in courses through honors thesis or undergraduate research projects, internships, or summer job experiences. One opportunity for undergraduates to gain experience in a research setting is through the Research Experience for Undergraduates (REU) program of the National Science Foundation. In this program, students apply to work on active research projects being conducted at host institutions. The program is open to U.S. citizens or permanent residents, and students are granted stipends to cover living expenses.

Some university programs have formal requirements for internships or summer work experience, whereas others simply stress the need for this experience and help students find such opportunities. Most state and federal conservation agencies hire undergraduate students during the summer. In addition, universities with strong graduate fisheries programs commonly hire undergraduates as summer technicians to work with graduate research assistants. Undergraduate students will find these summer positions rewarding; they provide an opportunity to apply techniques learned in the classroom as well as hone skills and acquire new ones.

Students should also seek opportunities to gain experience through volunteering for various activities. Volunteer activities can take many forms. For example, many state conservation agencies have field spawning operations for fishes such as walleye, northern pike, striped bass, trout, and salmon. Volunteering to work at such an operation can provide valuable experience in fish culture, fisheries management, and public relations, and also provides an opportunity to interact with working professionals. Students can also look to the graduate research program at a university to provide opportunity for volunteer work experience. Students might volunteer to gain experience in fish sampling

Box 1.1. The Hutton Junior Fisheries Biology program: an opportunity for high school students to obtain knowledge and skills.

The Hutton Junior Fisheries Biology Program is a summer mentoring program for high school students that is sponsored by the American Fisheries Society (AFS). The program was named in honor of the late Dr. Robert F. Hutton. Dr. Hutton was the first Executive Director of AFS (1965–1972) and AFS President in 1976–1977. In addition to his significant contributions to AFS, Dr. Hutton was a renowned fisheries administrator with the National Marine Fisheries Service and was known for his support of youth education as a dedicated member of the National Conservation Committee of the Boy Scouts of America.

The principal goal of the Hutton Program is to stimulate interest in careers in fisheries science and management among groups underrepresented in the fisheries profession, including minorities and women. Application to the program is open to all sophomore, junior, and senior high school students regardless of race, creed, or gender. Because the principal goal of the program is to increase diversity within the fisheries profession, preference is given to qualified women and minority applicants. Students selected for the program are matched with a professional mentor in their area for a summer-long, hands-on experience in a marine or freshwater setting. Assignments are made with participating organizations within reasonable commuting distance from the students. During the summer, students work alongside their mentors, collecting samples and analyzing data. A scholarship is provided to students accepted into the program (in 2006, the scholarship was for $3,000).

Since its inception in 2001, participation in the Hutton Program has grown from 23 students in the first class to 63 students in 2005. Increased participation has been made possible through the growing financial support provided by a broad array of agencies and individuals. In 2005, supporters included the National Fish and Wildlife Foundation (including funds from the National Oceanic and Atmospheric Administration [NOAA], the Bureau of Land Management, the U.S. Fish and Wildlife Service [USFWS], and the U.S. Forest Service [USFS]), NOAA Fisheries, the Alaska Department of Fish and Game, the North Carolina Wildlife Resources Commission, and the Wisconsin Department of Natural Resources. Units of AFS that provided support for the program included the Education Section, the North Central and Northeastern Divisions, and the Minnesota and New York Chapters. For more information and application forms, visit the Hutton Program section of the AFS web site (http://www.fisheries.org/html/hutton.shtml).

techniques, fish marking, care and handling of captive fishes, and end-less other possibilities.

Finally, university marine laboratories and biological field stations offer opportunities to gain additional knowledge and experience. This knowledge and experience may be gained through formal coursework at the laboratory or field station or through independent study. Addi-tionally, because field stations and marine laboratories are host to vig-orous field-oriented research programs, there are frequent opportuni-ties to gain experience as a paid or volunteer research technician.

Internships, summer jobs, and volunteer positions offer undergradu-ates the opportunity to experience various aspects of the profession first-hand and thus learn whether it really is the career they desire. The different employment settings—academic, agency, nongovernmental organization, and so forth—will provide students with the unique fla-vor of various fisheries employment opportunities. Furthermore, these experiences will help students establish professional contacts with po-tential employers and help identify graduate study opportunities.

Professionalism

While professionalism is thoroughly discussed in Chapter 15, we want to stress the importance of professional activities to undergradu-ate students. For example, the sooner that students become involved in professional societies, the more likely they are to gain employment and eventually contribute to the fisheries profession. In addition, life-long learning will be essential for a successful career. The learning process continues long after your formal education ends, and profes-sional societies can provide continuing education opportunities.

For many students, undergraduate club organizations provide the first opportunity to become professionally active. These clubs have a variety of names and a variety of affiliations. Some are affiliated with AFS, The Wildlife Society, or other professional societies, and some function effectively without affiliation. They may have a specific focus, such as fisheries, or may be broadly based, such as a biology club. These clubs provide an important opportunity for undergraduates to perform in service and leadership roles. Students may first serve simply as a member of a committee. Given continued involvement, students will likely have the opportunity to progress in their responsibilities, perhaps to committee chair or to an elective office. Such involvement helps students learn organizational and communication skills that will be essential in their professional careers.

The American Fisheries Society offers numerous opportunities for students to become professionally involved. The American Fisheries Society is composed of many organizational units, including chapters, divisions, and sections. Each unit functions as a subordinate to the international parent society, which has its own organization, programs, and governance. Chapters represent the most local level of organization and are usually delineated by state, university, locality, or region. Divisions are geographic units of the parent society and include the Northeastern, North Central, Western, and Southern Divisions. Sections represent various disciplines, including Early Life History, Education, Estuaries, Fish Health, Fisheries Management, Genetics, International Fisheries, Marine Fisheries, Socioeconomics, and many others.

There are a few AFS chapters that are directly affiliated with university programs. Many other university programs have a formal or informal relationship with their state, local, or regional AFS chapter. Participation in the professional and social activities in such programs will be enjoyable and rewarding. However, student members of AFS need to realize that they have all of the rights and responsibilities of any other member. Students are often unsure of their role in professional societies, but in AFS student members have one vote on important issues just as do all other members. A particular strength of AFS is its focus on students, and student involvement at all levels (chapter, division, section, and parent society) is common and encouraged.

At most levels of AFS, both undergraduate and graduate student members are eligible for awards to help them fund their travel to meetings. For example, each year the John E. Skinner Memorial Fund

(Skinner Award) supports the travel of deserving graduate students and exceptional undergraduate students to attend the AFS Annual Meeting. Other awards are available through various divisions, sections, and chapters. Attending local scientific meetings, such as Academy of Science meetings, can also be valuable experiences. University department chairs and department heads recognize the value of professionalism and will often provide vehicles to take students to professional meetings that are held within a reasonable distance of campus. An active student club often can facilitate such travel.

Sources of Additional Information

The web site of the American Fisheries Society (http://www.fisheries.org) contains a lot of information for students interested in a career in fisheries. For example, additional information about criteria for evaluating university fisheries programs is provided in the *Certification* section of the web site. The Education Section of AFS also maintains a separate web site focused on the educational requirements of fisheries professionals, which includes listings of academic institutions with fisheries and related programs (http://www.fisheries.org/education/programs.htm). Information about the REU program of the National Science Foundation, including instructions for how to apply, can be found on the program's web site (http://www.nsf.gov/funding/pgm_summ.jsp?pims_id=5517). Most fisheries-related opportunities are listed under *Ocean Sciences* or *Biological Sciences*. Lastly, the annual *Aquaculture Buyer's Guide*, published by Aquaculture Magazine (http://www.aquaculturemag.com), provides information on schools, coursework, and degrees offered for aquaculture programs.

Chapter 2

Developing an Effective Resume or Curriculum Vitae

TRENT M. SUTTON, REBECCA A. ZEIBER, AND EMMANUEL A. FRIMPONG

Whether you are applying for a summer job to gain professional experience, graduate school to further your education, or a permanent position in management, research, or academia, you will need to submit as part of your application package either a resume or curriculum vitae (CV). These documents serve as important marketing tools that will give a self-portrait or advertisement of you and will present your relative strengths, skills, and experiences to a potential employer. An effective resume or CV will provide an employer with an overview of who you are as a student or young professional, what you know and can do in relation to the position of interest, and what relevant skills, traits, and accomplishments you have achieved at this point in your education or career. Therefore, the objective of your resume or CV is to catch the eye of a prospective employer and secure an interview. Because employers may have to sift through dozens of these documents for a particular position, it is important that your resume or CV makes a strong first impression. As a result, these documents must be visually appealing, well organized, error-free, and written in a highly literate fashion. It is also important to update your resume or CV frequently in order to accurately reflect your most recent and relevant experiences and accomplishments.

A dilemma that you might encounter when deciding to apply for a position is what to submit—a resume or a CV. Therefore, it

Table 2.1. Comparison of elements that are typically included in a resume versus a curriculum vitae (CV).

Content Categories	Resume	CV
Contact Information	X	X
Employment Objectives	X	
Educational Background	X	X
Professional Experience	X	X
Professional Affiliations and Memberships	X	X
Professional Service	X	X
Honors and Awards	X	X
Grantsmanship		X
Teaching Experience and Interests		X
Publications	X	X
Presentations	X	X
Reviews of Journal Articles, Books, and Book Chapters		X
Student Mentoring		X
Experience and Certifications	X	X
Interests and Activities	X	
Professional References	X	

is important to distinguish between these two types of documents and identify when it is most appropriate to use one versus the other. While there are many similarities between a resume and a CV, there are also significant differences that serve to set these two documents apart (Table 2.1). A resume is a brief (two pages or less) summary of your professional qualifications, education, and experiences that allows a potential employer to quickly determine if you meet their job specifications (Box 2.1). In contrast, a CV includes these same components but in much greater detail. Further, a CV is typically longer (two or more pages), more formal in format, and provides a detailed listing of your professional service and activities, honors and awards, grantsmanship, publications, presentations, and other items of professional relevance (Box 2.2). In most cases, a resume is used when applying for temporary, seasonal or longer-term jobs, while a CV is used when applying for positions in academia or ones that are primarily research in nature. A CV is also used when applying for fellowships or grants. While this is the general rule of thumb in the United States and Canada, a CV that

Box 2.1. Example of a resume.

JANE M. DOE
123 Main Street
Tamaqua, Pennsylvania 17992
Phone: (215) 123-4567; Email: jdoe@localnet.com

EMPLOYMENT OBJECTIVE
Obtain a position as a fisheries technician to develop experience leading to a graduate position

EDUCATION
08/99-05/04   Keystone State University, Department of Natural Resources
Bachelor of Science, Fisheries and Aquatic Sciences
Cumulative GPA: 3.56/4.00

RELEVANT COURSEWORK
Aquatic Ecology, Aquatic Pollution, Fisheries Management, Ichthyology, Freshwater Entomology, Wetland Conservation, Fish Population Dynamics

EMPLOYMENT EXPERIENCE
06/04-Present   Pennsylvania Environmental Services, Harrisburg, Pennsylvania
Natural Resources Biologist I; J. McMahon, Chief of Fisheries Management Unit
· Assisted with backpack electroshocking of streams, identified fish species
· Collected routine and storm water samples from streams, recorded cross-section velocity data, calculated discharge rates, filtered samples for nutrient analyses
· Assisted with the collection and identification of anadromous fishes

05/03-09/03   Northern Lakes National Park, Grand Park, Minnesota
Biological Science Technician; T. Evans, Fishery Biologist, National Park Service
· Removed otoliths, stomachs, and scales from fish for further analyses
· Assisted with collections and radio telemetry of lake sturgeon
· Trained in CPR/First Aid, DOI's Motorboat Operator Certification Course

Box 2.1. Continued.

01/02-12/02	Keystone State Forestry Research Laboratory, Harrisburg, Pennsylvania
	Independent Study; R. Tyson, Assistant Professor of Watershed Management
	• Conducted literature reviews for macroinvertebrate sampling methods
	• Classified various orders of stream and wetland macroinvertebrates

PROFESSIONAL SERVICE

00-04	Keystone State University Student Subunit of the American Fisheries Society: Fundraising Committee, Chair (2003)
03-Present	American Fisheries Society, Member

COMPUTER SKILLS

ArcView GIS, SAS, SigmaPlot, Microsoft Word, Excel, PowerPoint, Publisher

includes detailed personal information is frequently required for jobs in Europe, Asia, and Africa.

As a final step in the process of submitting your application materials, be sure to include a cover letter with your resume or CV that has been customized to the specific position for which you are applying. The objective of the cover letter is to provide a brief overview of your experiences, skills, and accomplishments to a potential employer. While there will be some redundancy between your cover letter and resume or CV, the cover letter should highlight your key experiences and serve as a claim that demonstrates to a potential employer that you can meet the demands of the position in question. A typical cover letter is one to two pages in length, and is composed of three to five paragraphs that include the following information: (1) a self-introduction; (2) an argument or rationale for the strong connection between what you have to offer as a potential employee and what the employer needs in order to fill the position responsibilities; and (3) a closing statement with your up-to-date contact information (Box 2.3). Note that cover letters that simply provide a summarized review of the resume or CV are redundant with those documents and will be weak in comparison to those

Box 2.2. Example of a curriculum vitae (CV).

JANE M. DOE
123 Main Street
Tamaqua, Pennsylvania 17992
Phone: (215) 123-4567; Email: jdoe@localnet.com

EDUCATION
2005-Present. Master of Science (expected graduation: May 2006),
Department of Fisheries and Wildlife (Fisheries Science), Upper
Peninsula State University; Cumulative GPA: 3.89/4.00; Thesis title:
*Population dynamics and stock structure of walleye in Lake Gogebic,
Michigan*; Advisor: Dr. Julia Y. Hoops
1999-2004. Bachelor of Science, Department of Natural Resources
(Fisheries and Aquatic Sciences), Keystone State University;
Cumulative GPA: 3.56/4.00

PROFESSIONAL EXPERIENCE
Natural Resources Biologist I – May 2004 to August 2004. Pennsylvania
Environmental Services, Harrisburg, Pennsylvania.
Biological Science Technician – May 2003 to September 2003. Northern
Lakes National Park, Grand Park, Minnesota.
Independent Study Student – January 2002 to December 2002. Keystone
State Forestry Research Laboratory, Harrisburg, Pennsylvania.

PEER-REVIEWED PUBLICATIONS
Doe, J. M., J. Y. Hoops, R. A. Zoe, and T. M. Scout. In preparation. Impacts
of harvest on walleye recruitment in Lake Gogebic, Michigan. North
American Journal of Fisheries Management.
Johnson, R. L., **J. M. Doe**, and T. R. Donaldson. In review. Movement
patterns of lake sturgeon in northern Minnesota lakes. Journal of
Aquatic Biology.
Doe, J. M., N. A. Reily, and G. M. North. 2004. Potential effects of water
quality on rainbow trout populations in central Pennsylvania.
Pennsylvania Academy of Sciences 32:1603-1620.

TECHNICAL REPORTS
Doe, J. M. and T. R. Donaldson. 2004. Assessment of remnant lake sturgeon
populations in the lower Placid River, Minnesota. Minnesota
Department of Natural Resources, New Lake, Minnesota.

POPULAR PUBLICATIONS
Doe, J. M. 2001. Rapid bioassessment as a method of determining aquatic
health. Keystone State Daily Courier.

Box 2.2. Continued.

Doe, J. M. 2000. Brook trout stocking: pros and cons. Keystone State
  Aquatic Sciences.
Doe, J. M. 2000. Water quality concerns in our state. Keystone State
  Aquatic Sciences.

MEETING PRESENTATIONS
Doe, J. M., J. Y. Hoops, R. A. Zoe, and T. M. Scout. 2006. Changes in
  population structure of walleye in Lake Gogebic, Michigan.
  Contributed Poster. Annual Meeting of the Michigan Chapter of the
  American Fisheries Society, March 2006, Lansing, Michigan.
Doe, J. M., J. Y. Hoops, R. A. Zoe, and T. M. Scout. 2005. Impacts of harvest
  on walleye recruitment in Lake Gogebic, Michigan. Contributed Paper.
  Midwest Fish and Wildlife Conference, December 2005, Grand Rapids,
  Michigan.
Doe, J. M., T. R. Donaldson, and R. L. Johnson. 2005. Movement patterns
  of lake sturgeon in northern Minnesota lakes. Contributed Poster.
  Annual Meeting of the American Fisheries Society, September 2005,
  Anchorage, Alaska.
Doe, J. M. and N. A. Reily. 2004. Water quality as a factor in brook trout
  populations in three Pennsylvania streams. Contributed Paper. Spring
  Meeting of the Pennsylvania Chapter of the American Fisheries
  Society, March 2004, State College, Pennsylvania.
Reily, N. A. and **J. M. Doe.** 2004. Watershed land usage impacts on
  Pennsylvania brook trout populations. Contributed Paper. Spring
  Meeting of the Pennsylvania Chapter of the American Fisheries
  Society, March 2004, State College, Pennsylvania.

PROFESSIONAL AFFILIATIONS
American Fisheries Society
          National Chapter (2003-present)
                    Education Section, Member (2004-present)
                    Education Section Newsletter, Co-Editor (2004-present)
          Michigan Chapter (2005-present)
          Pennsylvania Chapter (2001-2004)
                    Continuing Education Committee, Co-Chair (2004-
                    present)
          Keystone State University Student Subunit (2000-2004)
                    Fundraising Committee, Chair (2003)
National Audubon Society (1998-present)
The Nature Conservancy (2000-present)

Box 2.2. Continued.

PROFESSIONAL CERTIFICATIONS
Department of the Interior's Motorboat Operator Certification Course, May 2002.
CPR/First Aid, June 2002.

HONORS AND AWARDS
John G. Gideon Memorial Award for Outstanding Fisheries Research, Michigan Chapter of the American Fisheries Society, 2005.
Keystone State Student Government Travel Grant, 2004.
Best Returning Member Award, Keystone State University Student Subunit of the American Fisheries Society, 2003.

VOLUNTEER EXPERIENCES
Assisted with Harper County Junior Envirothon at Mountain Top, Pennsylvania
Taught leadership and cooperation skills to 4H youths at retreats and camps throughout central Pennsylvania, 1999-2001.

---

that convincingly claim that there is a match between you as a potential employee and the employer. Like the resume or CV, the cover letter must also be well written, grammatically correct, and visually appealing. Anything less may dissuade a potential employer from considering you for the position in question.

When submitting materials like a cover letter via postal mail or email, be sure to use formal wording and the correct prefix for the person to whom the letter is addressed. Starting the letter with "Dear John Smith" or "Hello Ms. Jones" comes across as too casual and may jeopardize the chances of your application making it to the review process. Instead, start your cover letter with "Dr. Smith" or "Ms. Jones." In addition, make sure Dr. Smith and Ms. Jones really are the correct prefixes. Nothing is more embarrassing than addressing a letter to a male using a female prefix, and addressing a professor as "Mr." or "Mrs." instead of "Dr." can be equally detrimental to your application review. Attention to detail in your cover letter is of utmost importance. At its best, a cover letter tells the reviewer that your application is worth a look. Take time to ensure that yours is communicating what you want effectively and allows for a smooth segue into your resume or CV.

Box 2.3. Example of a cover letter to accompany a resume or CV.

20 January 2006

Dr. John A. Smith
Wisconsin State University
Department of Natural and Renewable Resources
555 Underwood Avenue
Hilltop, Wisconsin 56565

Dr. Smith:

I am interested in applying for the fisheries project leader position as advertised on the American Fisheries Society website. Please find enclosed a copy of my curriculum vitae, a list of references, and college transcripts as requested in the position announcement. My educational background in applied fisheries biology and management, coupled with my experience as a research technician and graduate student, has prepared me well to fulfill the responsibilities of this position. I appreciate the opportunity to provide you with an overview of my accomplishments in this letter.

I am a 2004 graduate of Keystone State University, where I received a B.S. in fisheries and aquatic sciences. Currently, I am a master's degree candidate in fisheries science at Upper Peninsula State University and intend to complete my thesis and defense by June 2006. During my academic career, my studies focused on freshwater systems, with a particular emphasis on fish populations in lakes at northern climates. I have significant experience using fish sampling gears such as seines, trawls, electrofishing, gill nets, and trap nets. Specifically, I have worked with lake sturgeon in Minnesota where I inserted radio tags and tracked adult fish. For my Master's degree research, I am evaluating the impacts of recreational harvest on walleye in Lake Gogebic, Michigan, which includes sampling fish using a variety of gears and conducting a creel survey of lake anglers. I also have experience operating small boats in lakes and bays, and last year I passed the Department of the Interior's Motorboat Operator Certification Course. In addition, I work well both independently and in a team, and have supervised crews as large as four people on extended sampling trips. Overall, I have a strong work ethic and am genuinely committed to a career in the fisheries field.

Based on my experiences and interests, I believe that I would be a strong candidate for the fisheries project leader position in the Department of Natural and Renewable Resources at Wisconsin State University. I would like to thank you for your time and consideration in reviewing my application materials. If you have any additional questions, please contact me directly via phone from 9 a.m. until 5 p.m. at (215) 123-4567 or via email at jdoe@localnet.com.

Sincerely,

Jane M. Doe
Research Assistant, Fisheries Biology

Getting Started on Your Resume or Curriculum Vitae

The keys to developing an effective resume or CV are to provide relevant information for the position of interest and to present that material in a well organized and error-free manner. As a result, you do not want to submit a document that lacks critical information or contains excessive detail, is written poorly or contains grammatical or spelling mistakes, or does not set the proper tone or context. Although there is no single standard format for developing your resume or CV, an effective document will be one that emphasizes your skills and traits that are most important for the position of interest and conforms to standard conventions within your particular discipline. Remember that first impressions are important, which can make a difference between securing an interview and having to continue your search for a position.

Regardless of whether you are developing a resume or CV, start by generating a list of all your background information organized into different categories. The first step in this process is to thoroughly and objectively analyze your professional experiences and education. For professional experience, note all your previous employment responsibilities for each position that you have held, including both daily routine tasks and any activities that necessitated special training and skills. Note that this category should not only include paid positions, but any volunteer work, internships, or work-study positions. If you have limited work experience, then the education category may be the primary focal area on your resume or CV. Be sure to include all colleges and dates that you attended them, degrees and dates that they were awarded, major and minor subject areas, relevant courses that were taken, grade point averages and academic honors, and involvement in undergraduate or graduate research. If you are developing a CV, be sure to also include categories such as professional affiliations, service activities, honors and awards, publications, presentations, grantsmanship, teaching experience and interests, journal article or book reviews, student mentoring, and relevant professional certifications.

As part of this information gathering process, be sure to develop a list of relevant personal skills and attributes. Keep in mind that employers will give greater preference to individuals with transferable skills that can be used in various work settings. For example, transferable skills include data analysis, problem solving, critical thinking, research and planning, decision making, teamwork, organization, management, leadership, communication (both verbal and written), and human relations. These are skills that you may have developed over the course of your professional and educational experiences, as well as during service activities such as serving as a committee chair or officer in a professional society.

Appearance of Your Resume or Curriculum Vitae

Organization and Format

Although the information that is included in your resume or CV will depend on the position for which you apply, be consistent in your organizational layout so that your document is easy for a potential employer to scan or read. The first consideration is to determine whether to organize your document following a chronological or functional format. While either format is acceptable, chronologically organized resumes or CVs are most common. However, it is common to organize the cover letter in a functional format to highlight major accomplishments.

A resume or CV organized following a chronological format provides a complete listing of all relevant activities or achievements within each category, usually listed from most to least recent (i.e., reverse chronological order). In contrast, a resume or CV that follows a functional format organizes information from many experiences around a functional area (i.e., teaching, research, service, administration, etc.). The final choice on which format to use is typically based on the student's level of experience. For example, if you have limited professional experience, you will most likely find that a chronological format best suits your needs for organizing the resume or CV. If you have greater professional experience, you will more likely utilize a functional format. Always keep in mind, however, that the type of position you seek may also dictate your choice of using a chronological or functional format.

An additional consideration for developing an effective resume or CV is to use gapping and parallelism. Gapping refers to the use of incomplete sentences to present information as clearly and concisely as possible. For example, you might write "Research Associate (May 2001 to June 2002): supervised technicians; conducted field work; completed data analyses; wrote final project reports." Through the use of incomplete sentences, you will eliminate unnecessary wording and allow a potential employer to quickly identify what you have been doing.

Parallelism refers to keeping the structure of your phrases and sentences consistent throughout the entire resume or CV. For example, if you use particular verb phrases in one part of the document, use these same word phrases throughout the entire resume or CV. Parallelism is particularly important within each category in order not to confuse a potential employer. In using both gapping and parallelism to develop your resume or CV, the guiding principle should be to increase the conciseness of the document as well as the ease of readability.

Customization

For both a resume and CV, different versions of these documents will be required for different types of positions. Keep in mind that length is not the determinant of a successful document. While it is important to present all the relevant information for a particular position, this information must be presented in as concise a manner as possible. Typically, resumes range from one page for an entry-level position or for a person with limited professional experience to two pages for more advanced positions or individuals with greater experience. In contrast, CVs are longer documents (two or more pages), with the length determined by the experience and qualifications of the person submitting the document. It should be noted that there are occasions, typically for fellowship or grant applications, where shorter CVs (one or two pages) are required. In these instances, the required categories and information to include are well defined by the organization requesting the document.

Writing Style

While a resume or CV that is well written and free from errors will not guarantee that you will get an interview, a single mistake may result in the outright rejection of your application materials. As a result, make sure that your document is completely free from typographical errors and does not contain any flaws in grammar, punctuation, or spelling. To produce a watertight resume or CV, be sure to critically and objectively review it multiple times. In addition, have a friend, fellow student, professor, or academic advisor review it as well in order to provide a critical evaluation prior to submission.

In writing your resume or CV, be sure that you write in the third person and avoid using the pronoun *I*. In addition, use an active voice throughout the document and only employ action verbs. The following list of action verbs are typically used in resumes or CVs:

Administered	Analyzed	Collected	Completed
Communicated	Conducted	Coordinated	Created
Designed	Developed	Directed	Evaluated
Increased	Led	Managed	Modeled
Observed	Operated	Organized	Oriented
Prepared	Researched	Sampled	Started
Supervised	Taught	Trained	Wrote

Production of a Resume

The appearance of your resume or CV must be both high in quality and well organized. In all cases, the presented information must not be crowded or cluttered. When possible, balance the layout of your resume or CV to enhance its appearance and readability. As a general rule of thumb, use one-inch margins whenever possible and never use margins smaller than 0.75 in. Between each section, be sure to leave a sufficient amount of white or empty space in order to adequately separate the categories. For section headings, use **UPPERCASE BOLD LETTERS** and a combination of **Uppercase/Lowercase Bold Letters** to highlight subsections. In addition, use variable but consistent font type and sizes (between 10 and 14 point) for section and subsection headers to aid in distinguishing categories. Commonly used and appealing font types include Arial, Times New Roman, Helvetica, Courier, or CG Times.

To produce a visually appealing resume or CV, use an appropriate type of paper and ink. For example, use white, off-white, ivory, or light gray paper with black or navy ink to provide high contrast. Do not use bright or gaudy colored paper and ink, as that will detract from the quality of your document. Instead of using typical photocopy or printing paper, purchase high-quality paper from an office supply store that is 20-50 pound bond, 100% cotton fiber, and is standard in size (i.e., 8.5 x 11 in). In addition, do not use photos from your latest fishing trip on the background of your paper or stamp fish drawings on any part of your document. These items are too informal to include in an application package and make the applicant appear immature. Print your resume or CV using a high-quality laser printer and only print your document on one side of the paper (never front-and-back). Use the same printer, paper, and ink for printing your cover letter. While you should staple your resume or CV, use a paper clip to attach it to the cover letter. Finally, do not fold your resume or CV; instead, send it to your potential employer in a 9 x 12 inch envelope.

Categories and Content for a Resume or Curriculum Vitae

Header/Personal Information

As stated previously, the point of a resume or CV is to make a statement about you, grab the attention of a potential employer, and land you an interview. The best way to facilitate this process is to begin with your name at the top of the page, not "Resume" or "Curriculum Vitae." One glance through the document will reveal its nature, so using such a title is redundant. List your name at the top and center of the document in a slightly larger, bold font. Do not include middle initials or suffixes (i.e., Jr., Sr., III) unless you always add them when signing your name. Prefixes like "Mr." or "Ms." are typically avoided as they will add unnecessary clutter to an otherwise clean header at the top of your resume or CV.

Your address is generally listed under your name, followed by a phone number where you can be reached. If you choose to list your office phone number, be aware that your current employer may not be keen on the idea of you spending time discussing another position or interviewing over the phone during your workday. A good alternative might be to allow the reviewer to initially contact you at work, and then set up a time outside of work to discuss your qualifications.

Lastly, in this age of technology, it is necessary to provide an email address along with your other personal information. Potential employers sometimes make the first contact with candidates via email to set up an interview time and date, so do not omit this important detail from your resume or CV. If you choose to list your work email address, this will increase the odds that your supervisor finds out that you are looking for another position. In addition, some companies track emails sent by their employees, and many restrict email use to company business only. Therefore, it might be wise to list a personal email address rather than a business one to ensure this contact is not interfering with your current employment. However, use an email address that is appropriate for this type of communication (e.g., janedoe@dishnet.com) and is not too casual (e.g., fishlover@dishnet.com). In general, this first section of your resume or CV is meant to provide a potential employer with the necessary information to contact you. It should be accurate, easy to understand, and stand out without overwhelming the rest of your document.

Employment Objectives

An employment objective placed near the top of your resume can create a smooth transition to the rest of the document (note that employment objectives are not included on a CV). In addition, an objective statement gives your resume a sense of direction and a brief glimpse into your future goals. If you choose to include an objective, it should be clearly written and concise. Typically, a one-sentence objective is adequate, although it could be expanded to a short paragraph if space permits. It should focus on your short- and long-term goals and be somewhat general in nature. A narrow objective can be too constricting, and could potentially exclude you from a position for which you may be qualified. Similarly, an objective that is too general appears vague and does not lend focus to your resume. When stating your goals, do not list position titles, such as "Entry-level fisheries position with a state agency." Instead, include a sentence or two on specific skills you have or tasks you would like to perform and how they fit into your goals. Your objective statement may also change depending on your audience. For example, it may be inappropriate to send a resume for a laboratory technician position with an objective stating your interest in conducting field work on a boat every day. By tailoring your objective for each position, you can help prevent your resume from moving to the bottom of the pile. Although it is not necessary to include an objective section, it may serve to set the tone of your resume and can lend an insight into your strengths as a potential employee.

Educational Background

The first major section of a resume or CV should highlight your educational background. Typically, you should list this information in reverse chronological order, with your most recent degree listed first. Include your graduation date, or expected graduation date if forthcoming, institution name and location, major/minor, and dates attended. For M.S. and Ph.D. degrees, include your dissertation or thesis title and name of your advisor. If you attended more than one institution, list every school attended regardless of your graduation status. Other forms of education, such as an Associate's degree, military education, and technology courses, should also be included in this section. Unless you only have a Bachelor's degree, do not include your high school name and location because most employers do not need this information. If you would like to highlight your grade point average (GPA), do so for each degree earned and include the scale (i.e., 3.56 on a 4.0 scale, or 3.56/4.0). If your cumulative GPA is low, you should highlight your GPA in your major as well to show employers that you excelled in the subjects that match your current career choice. Although it is not necessary to list the courses you took while in school, you may want to include a short list if you majored in something other than fisheries. For example, if you earned a Bachelor's degree in environmental sciences and are currently seeking a fisheries position, it might be wise to highlight the statistics, watershed management, and freshwater ecology courses you took to demonstrate to employers your interest in and knowledge of a related field. This coursework could either be listed at the end of the education section or in a separate section of their own, depending on space limitations and the amount of information you want to convey. As your career progresses, you may find it unnecessary to list courses because your work experience may carry more weight with employers reviewing your resume or CV. Lastly, the education section of your document should provide a nice segue into the following work experience component.

Professional Experience

For many people, the professional experience section of their resume or CV is the longest and most important portion that highlights their skills and capabilities in the work place. Similar to the education section, your professional experiences should be listed in reverse chronological order, highlighting your most recent job first. If your most recent job is not applicable, then draw attention to your more relevant positions by placing them near the top of this section. You may want to tailor your resume or CV for the particular job to which you are applying if you have a wide variety of experiences. Highlight the professional experiences that show employers you are the right person for the position. If you are just getting started in your career and have relatively few relevant experiences, list a few non-professional positions that show your capabilities and work ethic.

Begin each position description with the dates that you worked there. You should either write out the entire date (i.e., February 1996 – December 1999) or abbreviate it numerically (i.e., 02/96 – 12/99), but be consistent throughout your resume or CV. List your position title, name and location of employer (city and state will suffice), and a brief description of your responsibilities. Generally, a resume focuses on these descriptions, while a CV might only list the job title and employer. As your list of experiences grows, it will become challenging to find space to explain your work responsibilities. Therefore, many applicants who have significant professional experience will submit CVs without much explanation for past positions. If you do include a job description, you may wish to use either gapping or parallelism (see *Organization and Format*). Whatever form you choose to highlight your experiences in, keep it consistent throughout the section to maintain a good reading flow. The key here is to say as much as you can about your experiences without overwhelming your audience. You may choose to focus on skills and accomplishments instead of just your responsibilities at that position, including any special recognition you received.

For previous positions, do not list your salary or reasons for leaving unless specifically asked to do so. Employers often save those questions for the interview, so it is unnecessary to add clutter to your resume or CV with this information. By excluding these details, it allows you the opportunity to explain the situation to the employer directly.

Overall, the professional experience section of your resume or CV should demonstrate to potential employers that you have the necessary skills to make you the most qualified applicant for the position. Content and style play equally important roles in the impact this section can have on your audience. There is no point in having years of experience if you do not present that information in a clear, concise manner so that employers can easily understand your work history.

Other Skills and Accomplishments

This section of your resume or CV will provide you with an opportunity to list and describe additional skills and accomplishments to a potential employer that do not fit in any of the previous sections. The content and organization that you use will differ considerably depending on the stage of your career and the degree of importance you attribute to each category relative to the requirements of the position you seek. Consider consolidating headings if you do not have much information to include in each category. Because brevity is always important, only choose to highlight those skills and accomplishments that are directly relevant to the position for which you are applying.

Professional Memberships and Affiliations—Becoming a member of a professional organization or society provides you with an opportunity to develop the professional skills, philosophies, and value system which typifies that affiliation. Furthermore, these memberships will often be interpreted by a potential employer as an indicator of the level of enthusiasm that you have for your chosen area(s) of study and your recognition of the importance of belonging to a professional organization. Be sure to also include any memberships in student organizations or chapters, professional societies and associations, and other organizations with a focus on relevant natural resources issues (e.g., The Nature Conservancy, Desert Fishes Council, National Wildlife Federation). If you decide to include professional honoraries, be sure to provide the purpose of the organization as a potential employer may not be familiar with that organization. In all cases, you should not list affiliations with activist groups with parochial or radical interests on your resume or CV.

Professional Service—Becoming involved in professional organizations, academic institutions or places of employment, and student clubs provides you with an opportunity to develop leadership skills and professionalism. As a result, these types of positions should be listed and highlighted on your resume or CV. Examples of officerships include serving as the president or vice president (or president-elect), secretary, treasurer, chair of a committee, or membership and participation on a committee within a professional society or student club. In some cases, it will be useful for you to provide a brief description of your responsibilities for each position so that a potential employer can more easily determine what you might have accomplished in this capacity. Because attributes such as leadership, professionalism, responsibility, and maturity are valuable for any position that you may seek, it is important that you make a potential employer aware that you have developed such traits.

Honors and Awards—The honors and awards that you receive can show employers that you have the extra motivation and talent they are seeking. Academic and non-academic honors and awards should be listed under separate headers if you have more than just a couple in each category. However, if your award has a significant financial component, it may be more appropriate to list it in a section for grantsmanship (see next section). In general, honors received during high school should be left off of the resume or CV unless you have not yet received your Bachelor's degree.

The types of honors and awards that you should include on a resume or CV can range from Dean's List recognition to athletic awards. Service awards, academic achievements, awards for manuscripts and presentations, and community involvement would all be appropriate to include in this section. A listing of each honor or award, as well as the date it was received and a brief explanation, will help to clarify any questions a potential employer might have regarding the nature of your achievement.

Grantsmanship—Involvement in the writing of grant proposals, securing project funding, and participation in other fundraising opportunities provide evidence of your productivity, leadership, initiative, and potential for future success. As a result, you should list all of these activities in this category. If you are a student or junior professional, you should also include in this section of your resume or CV any meeting or travel grants that you have received as well as any supplemental support or grants obtained for undergraduate or graduate research. If you are applying for a position in academia or a research position with a state, federal, or tribal natural resources agency, provide the following information for all grant proposals: year submitted and project duration, title of the grant, funding source, and amount of funds requested. In addition, indicate whether you are serving as the principal or co-principal investigator for each proposal and list any collaborators. If you have received a fellowship, particularly if it was awarded for your thesis or dissertation research, you should also include it in this section of your resume or CV.

Teaching Experience and Interests—Obtaining a wealth of information in your area of interest or specialization may not make a substantial contribution to your field unless you learn how to present those ideas to a larger audience. By developing teaching experience, you will demonstrate to a potential employer your ability and desire to share knowledge with others through direct classroom lecturing or outdoor laboratory activities. Furthermore, teaching requires you to develop organizational, leadership, and communication skills, all attributes which any potential employer will find desirable. In your resume or CV, be sure to include any classes that you have taught or served as a guest lecturer, laboratory or teaching assistant responsibilities, and instructional teaching and associated activities that you may have conducted in a field setting. You should also provide information on your teaching experience by school, the courses taught, their instructional level, and the dates that you taught the course(s). In academia, a potential employer will also want to know what undergraduate and graduate courses you are interested in and qualified to teach so that they can gauge your potential involvement within their curriculum.

Publications—Publishing research findings in technical journals or the popular media demonstrates to a potential employer that you can communicate effectively to an audience in a written format. When including publications that you have obtained, always list them in reverse chronological order on your resume or CV. General publication categories include peer-reviewed or refereed articles, books and book chapters, non-refereed publications, proceedings, technical reports, and popular articles. For the peer-reviewed article category, use the following subheaders to categorize your publications: published, in press, in revision, in review, in preparation. Note that a long list of publications in the "in preparation" category relative to the other subcategories may be viewed by a potential employer as an attempt to make your list of publications more impressive than it is in reality. If your list of publications is short or there are not many categories, include them all under a single header and identify the different categories. For all publications, use a standardized citation format such as the journal style followed by the American Fisheries Society. If you have won an award for a paper, be sure to indicate that honor in this section of your resume or CV as well as the section reserved for honors and awards.

Presentations—Giving oral presentations and posters at professional conferences, workshops, research symposia, and public meetings also demonstrates to a potential employer your ability and willingness to communicate to groups ranging in composition from professionals to lay people. You should list oral presentations and posters in a similar format, typically the same one used for publications. General presentation categories include technical presentations, non-technical presentations, and seminars. Within these general categories, you may wish to use subheaders to organize your presentations into oral and poster presentation groupings. In addition, you should also note which presentations were invited. For all presentations, include the names of all co-presenters, the date of the presentation, the title of your presentation, and the meeting and location where your presentation took place. Like your publications, presentations should be listed in reverse chronological order. Do not list your thesis or dissertation defense seminars in this category because these presentations are standard requirements at most colleges and universities.

Reviews of Journal Articles, Books, and Book Chapters—At some point in your career, you may be called upon to review journal articles, books, or book chapters that have been submitted for publication. This opportunity allows you to make a contribution to your area of specialty through the peer review process. Although these types of reviews are typically reserved for more established professionals, students may be called upon to review journal articles, as well as magazines, newspapers, and other popular publications. The latter reviews should be cited in the resume or CV following the same format as journal, book, and book chapter reviews.

Student Mentoring—In most professions, mentoring a student in your field of study shows that you have the ability and initiative to teach, guide, and encourage someone who is getting started in the profession. Mentoring should be a learning process for you as well. Furthermore, mentoring a student could be a chance for you to get a more in-depth perspective of a different aspect of your field of study. Being a tutor or advisor for a student in a certain subject area could also be included under a mentorship category. This section also includes service on graduate thesis or dissertation committees, as well as supervision of graduate and undergraduate students. Students are generally listed by year and level (M.S. or Ph.D.), and you should indicate whether you are a committee member or chair. It should be noted that graduate students should also list mentoring experiences if they have served in this capacity for undergraduate or high school students.

Experience and Certifications—It may benefit you to include a section in your resume or CV that lists your skills and training courses that did not fit into any other category. For instance, if you are SCUBA certified and are applying for a job near a lake or bay, an employer may place your CV at the top of the pile because of the usefulness of this skill for a particular position. Being trained in a particular skill or area may save the employer time from having to train someone else if you get the position. Certifications such as First Aid, CPR, motorboat operator certification, or fish identification will be useful for a career in fisheries and aquatic sciences, so consider receiving formal training in these areas if possible.

Computer skills are also an absolute necessity in this day and age, so be sure to list any experiences that you may have with different types of software packages. Programs like ArcGIS (Environmental Systems Research Institute, Redlands, California) or SAS (SAS Institute, Inc., Cary, North Carolina) may be an important aspect of any position type, and the more computer savvy that you can be, the better off you will be in your career. Always be sure to list all the relevant training in these capacities that you have accumulated as these skills may help you win an interview.

Interests and Activities—Many people choose to list their interests and activities outside of the workplace to demonstrate that they are well-rounded individuals. If there is space for a section like this on your resume or CV, it may benefit you to include a listing of your relevant interests. There have been occasions where employers have seen a listed activity that lends an insight into an applicant's personality or work ethic, and may be a factor in winning you an interview. Whenever possible, include those interests that mesh well with the position you are seeking. This listing is generally more acceptable in preparing a resume for an internship or entry-level position. As your resume expands into a CV, you may find the listing of interests and activities to be extraneous and not want to include them.

Professional References

Professional references should be just that - professional. It is usually unacceptable to have your relatives or friends listed as references for you unless the job announcement specifically asks for that. Instead, include the reference's name, title, name and location of employment, telephone number, and email address for three to five people who know your work capabilities. This list may include a past supervisor from a volunteer position, a boss who has worked closely with you at a permanent or temporary job, or an advisor or professor who knows your strengths and weaknesses as a researcher or student. By listing the names of people with whom you have worked directly, it allows employers to gain an insight into your work ethic instead of speaking with a reference who barely knows you. It is necessary to ask these people first if they are willing and able to serve as a reference for you. Imagine the reaction of your former employer who gets an unexpected call from your new employer regarding your work ethic— the fact that you didn't ask permission to put his or her name down as a reference might affect the type of reference you get.

Some resume guides suggest not listing references in a resume or CV and instead stating "References available upon request." However, in the fisheries field where networking and name association are important, your references may be the deciding factor on whether or not your resume gets a second look. Consider an employer who is looking at your resume and another applicant's resume, both with comparable education and work experience. If your resume includes the list of fisheries professionals that the employer knows personally, he or she might feel more comfortable relying on the word of that reference instead of an unknown reference listed in your competitor's resume. An alternative solution is to include a list of references on a page separate from your resume or CV. This page should also list your name and personal data similar to the boldface heading at the top of your resume or CV, and have between three and five references listed underneath. Be sure not to attach written recommendations or job evaluations unless specifically asked to do so. The point of your resume or CV is to be clear, concise, and uncluttered; including superfluous documents just adds to the paper pile.

Concluding Remarks

The goal of developing an effective resume or CV is to construct a comprehensive self-portrait of your abilities that secures you an interview. However, the process of developing the necessary skills and experiences that you will need to include in your resume or CV to make you competitive for a position begins long before you get to this point (see Chapter 1). The challenge for you in developing your resume or CV is to advertise your capabilities in a manner that best represents what you can bring to a position of interest.

In addition to providing a comprehensive overview of your relevant professional experience and accomplishments, it is important that the information you include in each of the different categories of your resume or CV be truthful. Never provide false information or embellish your credentials to impress a potential employer. If you do provide information that is untrue and this is found out by a potential employer, the consequences could prove to be severe. At the very least, your credibility as a professional will have become seriously damaged and this could make it difficult for you to secure a position in the future.

Make sure that your resume or CV is as flawless and well designed as possible. As stated before, review your document more than once and also have a friend, colleague, advisor, or professor provide a critical evaluation. This process will not only result in the development of a resume or CV that is free of errors, but it will also help to ensure accuracy of your information. Because this document along with the cover letter will be the sole means by which a potential employer will make a decision regarding whether or not to grant you an interview, it is critical that you carefully prepare your resume or CV.

At most colleges and universities, there are student writing laboratories or placement services that provide help sessions and workshops that can assist you with preparing your resume or CV. Near most college or university campuses are office supply stores or commercial printing or copying centers that can assist you with selecting the appropriate type of paper, ink, print style, and layout. In some cases, these companies will professionally typeset and print your resume or CV (for a fee), which will result in the production of a highly professional document. Be sure to take advantage of the various reference aids that are available in textbook form (see *Suggestions for Additional Reading*) or on the Internet. Finally, examine as many resumes or CVs as possible and incorporate the wide range of advice that is provided to you by your reviewers to tailor a document that best meets your needs for a particular position. While following the guidelines suggested in this chapter will not guarantee success in securing a job, they will allow you to develop an effective and professional resume or CV that should help to increase the likelihood that you will be considered for an interview.

Suggestions for Additional Reading

Anthony, R., and G. Rose. 1994. The curriculum vitae handbook. Rudi Publishing, Iowa City.

Heiberger, M. M., and J. M. Vick. 2001. The academic job search handbook, third edition. University of Pennsylvania Press, Philadelphia.

Jackson, A. L., and C. K. Geckeis. 2003. How to prepare your curriculum vitae. McGraw-Hill Company, New York.

McDaniels, C., and M. A. Knobloch. 1997. Developing a professional vita or resume. J. G. Ferguson Publishing, Chicago, Illinois.

Neal, J. E. 1996. Effective resume writing: a guide to successful employment, second edition. Neal Publications, Perrysburg, Ohio.

Chapter 3

Pursuing Graduate Studies in Fisheries

ALEXANDER V. ZALE

The simple truth is that a graduate degree is now required for consideration for most entry-level fisheries positions. Whereas position descriptions for many fisheries jobs do not explicitly require graduate degrees, most of the applicants for such positions will have such degrees, typically rendering applicants without graduate degrees uncompetitive. Therefore, admission to graduate school is a crucial step in succeeding in this profession. However, the process of getting into graduate school at most colleges and universities is quite different from applying for admission to an undergraduate program; merely sending in your application materials will not suffice in most cases. Nor will simply meeting the minimum requirements. The opportunities are much more limited and the selection process is rigorous. Often, just a handful of positions are available annually, even in major fisheries programs. The competition can be fierce. You should therefore think of getting into graduate school as getting your first real job, and, as in getting a job, you should be ready to prepare, search, and compete for a graduate opening (Zale et al. 2000).

Before You Apply

A number of questions require your careful consideration before you start the search and application processes. First, should you go to graduate school immediately after completing your undergraduate degree or should you work for a year or so first?

If you cannot define your career goals or reasons for going to graduate school, or your motivation is not strong, you should obtain work experience first (Orth and Adelman 1996); graduate school is a major commitment. An internship or technician position will hone your interests and enhance your qualifications. The only danger is that you may become less motivated to make the sacrifices required by graduate school after drawing a paycheck and enjoying the pace, schedule, and benefits of a real job. Some individuals never go back and therefore never achieve their full potential. Most, however, get frustrated by the limitations in advancement and responsibility incurred by the lack of an advanced degree and are thereby motivated to return to school. Commonly, they wish they had done so sooner.

Are you qualified for graduate school? Minimum requirements typically include an undergraduate degree in fisheries or a related discipline (e.g., biology, zoology, marine science, ecology, natural resources, wildlife, forestry) from a good college or university, a GPA above 2.8, and combined verbal and quantitative Graduate Record Examination (GRE) scores of about 1000. Of course, meeting these minima is insufficient if better applicants exist. Your chances are much better if your undergraduate GPA exceeds 3.5, your combined GRE score exceeds 1300, and you have work experience, great references, and have been an active member of a professional society, particularly the American Fisheries Society (AFS). Individuals with intermediate qualifications stand an intermediate chance of acceptance.

Should you apply only to schools that have named programs in fisheries and named degrees in fisheries? Limiting yourself to named fisheries programs will limit your opportunities greatly, as relatively few schools that offer graduate education in fisheries actually include the word "fisheries" in their program title. Many fisheries professionals did not receive their graduate education in named fisheries programs. Similarly, named fisheries degrees are relatively unimportant, though some state fisheries agencies rank job applicants with such degrees higher than those receiving other degrees (see Chapter 4).

Should you pursue a Master of Science (M.S.) or doctoral (Ph.D.) degree? The answer depends on your timeline, career goals, and abilities. An M.S. degree typically takes about 2.5 years, whereas the duration of a Ph.D. program can be 3.5 to 5 years. An M.S. degree is the norm for freshwater fisheries management positions. In fact, a Ph.D. degree may be viewed as a liability for a management position because doctoral graduates are perceived to be more interested in research and academic positions than in staying in management. This perception is less widespread among marine agencies. University research and teaching faculty positions require Ph.D. degrees, but even if you plan on eventually getting a Ph.D., you may still want to get an M.S. degree first. Relatively few students have the experience and ability to excel at the Ph.D. level without first getting an M.S. degree, though a fair number of individuals erroneously think they can. An M.S. degree provides experience in conducting research that typically enhances the quality and efficiency of a subsequent Ph.D. research program. Students who take the direct Ph.D. route often take a year (or more) longer to complete their degrees than students who have already completed an M.S. degree because of their inexperience, and the quality of their research and research productivity often fall short as well. For these reasons, I myself do not consider individuals without M.S. degrees for Ph.D. positions in my program. Moreover, Ph.D. graduates who published their M.S. research are more competitive for postdoctoral and faculty positions because they typically have more publications than Ph.D.-only graduates. Individuals with both degrees also tend to have a broader technical and geographic knowledge of the profession, especially if they acquired the degrees at different schools and studied different topics for each degree. The only advantage of the direct-Ph.D. route is a shorter time to completion of one's education, and that advantage (perhaps only 1 or 2 years in reality) may not be as great as it may at first appear. Carefully weigh this advantage against the disadvantages when making your decision.

How does graduate school differ from the undergraduate experience? Whereas coursework is the centerpiece of an undergraduate degree, the research process (including research issue identification, literature review, study design, proposal development, field work, experimentation, data analysis, writing, presenting, defending, and publishing) takes center stage in a graduate degree program (Rossman 1995; Fischer and King 1998). The purpose of graduate coursework is primarily to rectify deficiencies and gaps in one's undergraduate

program, especially in quantitative and statistical areas or to meet certification requirements, provide access to unique concepts and to faculty members that are otherwise unavailable, and also to take courses
directly related to one's research topic. On the other hand, some marine graduate schools require students to take a prescribed and heavy
set of courses during the first year of residency before the research
component of the degree can be initiated.

Graduate school is a year-round endeavor. Summers and breaks
are spent conducting research, and late nights are common. Field work
is often conducted in geographic isolation (Gibbons 1998) far from
friends and loved ones. Graduate students accept most or all of the
responsibility for the success of their projects, especially at the Ph.D.
level, and are expected to complete work as specified, meet deadlines,
file reports, maintain equipment and vehicles, ensure safety and regulatory requirements are met, and supervise technicians. The expectations, responsibilities, and workload are much more demanding than
at the undergraduate level; some students fail to meet the challenge
and wash out.

Finding a Graduate Student Position

Individuals with graduate degrees get jobs because they have ac-
crued valuable skills, experience, and professional development while
getting these degrees. Graduates who worked on actual fisheries management problems for their graduate research are therefore especially
competitive for fisheries management jobs, because their experience
prepares them explicitly for the work they will be expected to perform
on the job. The same goes for other fisheries subdisciplines (e.g., fish
health, human dimensions, population dynamics, ecology, aquaculture, statistics, physiology, toxicology, genetics, and early life history).
Therefore, it is advantageous to find a graduate research project that
addresses an issue in the subdiscipline that you expect to focus on in
your career and that makes a significant contribution to that subdiscipline. If you have not yet decided on an area of specialization, look
for a relevant, well-funded project at a quality school with a productive and respected advisor whose former students are successful. You

can decide on the specific direction of your career later as you gain familiarity with the profession.

Significant graduate research projects are rarely possible without appreciable resources such as vehicles, boats, equipment, laboratories, technicians, supplies, and agency support and cooperation. Faculty acquire funds to pay for such resources and the graduate students who perform the research by preparing and submitting grant proposals to funding agencies. When professors successfully secure grant money, they select the best students available to work with them on the research. Selected students are typically supported by research assistantships, which provide modest salaries for subsistence and tuition. Students and faculty find each other through one of three mechanisms.

The rarest approach involves the selection of graduate students from a pool of applicants who have applied to a program "blindly" (i.e., not to a specific professor or project). Few schools use this approach exclusively anymore, but many others still maintain applicant pools and occasionally select students from them, particularly when other approaches fail. Contact the programs you are interested in and determine if any use pools exclusively. If so, submit the required application materials before the specified deadlines and await a response. Some of the following advice may be useful in assembling a successful application. If a school does not use an applicant pool exclusively, the other techniques tend to be more effective, but applying to the pool likely would not hurt, except perhaps for the application fee.

Many preeminent faculty select graduate students from among those individuals who have previously contacted them about potential graduate opportunities. This technique is efficient and effective for highly qualified students (i.e., those with excellent grades and GRE scores, honors and awards, plenty of experience [especially undergraduate research, internships, work study, and summer technician positions, as well as experience in real jobs], and who have a track record of productive working relationships with supervisors, professors, and coworkers). When a professor receives grant funding, he or she may contact such prospective students to determine if they are still available and interested in the specific research. Clearly, this is a favorable position to be in, but success with this approach requires considerable preparation (see Chapter 1). Do well in school, get experience, and get along well with others. Start the search for a position early, about a year before you expect to enroll (Allen 1993). Study for the GREs and take them as soon as feasible, retaking them if your scores need improvement. Take the GRE biology subject exam; few schools require it anymore, but a high score is an impressive attribute. Go to professional meetings and talk to faculty members. Give people a chance to get to know about you, your plans, and your interests. Your advisor or supervisor will be glad to help with introductions. Do things that get you noticed such as actively participating in professional societies, serving on committees, volunteering for tough assignments, running for office, making presentations, applying for an AFS Skinner Memorial Travel Award, and publishing technical and popular articles. These are all indications that you are serious about this profession and have the energy and motivation to be productive. Finally, follow the directions below to make your prospective graduate advisor(s) aware of your interest in their program(s). When a position becomes available, faculty will already have a favorable impression of you and contact you.

Sometimes faculty do not already have someone in mind, or even if they do, are interested in seeing if better prospects might be available. To find out who is available, they send out "assistantship announcements" and hope that highly qualified potential students will see their announcement and apply for the vacancy. Copies of the announcements are circulated by email among professors, who forward the announcements to students that they know are looking for graduate opportunities. They may also post paper copies on a bulletin board or keep them in a special file. Ask your advisor where they are kept at your school. Paper copies are often also posted on temporary "jobs boards" at professional and scientific meetings. However, the most important and efficient medium available today for dissemination and retrieval of assistantship announcements is the Internet, especially the AFS Job Center Online (see *Sources of Additional Information* at the end of the chapter). Almost all advertised fisheries assistantship vacancies are posted there. Responding to an assistantship announcement can be especially effective because the professor is actively searching to fill a vacancy and you can focus your application materials on the specifics of the advertised position.

My focus here is on research assistantships because I believe that they offer the best opportunity for graduate education for the typical graduate student. Other common mechanisms are teaching assistantships (which require a student to teach to receive a stipend and therefore detract from time spent conducting research) and fellowships. Fellowships (e.g., National Science Foundation, Environmental Protection Agency STAR) are highly desirable because they involve no obligations other than research and a student's own teaching interests, better funding for publishing, equipment and stipends than assistantships, and freedom to explore one's interests that arise from an initial line of investigation. However, the competition for fellowships is fierce. Only Ph.D. and exceptional M.S. students receive them, and some applicants are unsuccessful in ever securing one. Other support is needed while waiting for a fellowship to be awarded. A research assistantship is therefore often a more dependable funding mechanism for the typical graduate student, because it already exists at the beginning of a student's graduate program, offers sufficient funding to complete the research and degree requirements, is available to exceptional and average students alike, and concerns a topic that both a funding agency and professor have decided is relevant.

Evaluating a Potential Position

You should be selective when choosing a university, advisor, or project, but be selective for the right reasons (Grossman 1998; Box 3.1). Try to get on a pertinent project at a quality school with a productive and respected professor whose former students have the types of jobs you desire. Make sure that the project is fully funded, including an assistantship that you can subsist on. Do *not* be overly selective about where that school is and what species of fish the project involves—you should be willing to go anywhere and work on any relevant topic in your chosen subdiscipline. You will find that any locale is tolerable for a few years if you are addressing a challenging problem, working with good people, have the resources you need to do good research, and are doing the kind of work that will help your

Box 3.1. Questions, organized by themes, that can be asked of faculty and past and present graduate students to investigate prospective graduate programs and advisors. Answers to some of the questions may also be found through internet research.

Questions for Faculty

    A.   Advisor-Graduate Student Relationship

- How many students do you usually have?
- How often do you meet with your students?
- Do you have an open-door policy with your students, or do you meet with them by appointment?
- Do you meet with your students one-on-one, as a group, or both?
- Do you provide funding to cover all of the resources that your students need to conduct their research (e.g., vehicle, computer, equipment, technicians)?
- What laboratory and office space is available for your students?
- Is funding available to support a student throughout the duration of this study?
- How long will it take to complete this degree?
- Do your students present and publish their work? If so, where and how often? Do you pay travel and publication costs?
- Do you encourage students to attend scientific meetings if they are not presenting, and do you fund their travel when they do?
- What are the backgrounds of your current and past students?
- What proportions of your students are in M.S. and Ph.D. programs?
- What proportion of your students fails to complete their degrees?
- What are your former students doing now?
- What are your expectations for your students with regard to work hours, time off, professional activities, and other employment?

    B.   The Graduate Program

- How many fisheries faculty are in the program?
- Do the students and faculty actively publish in peer-reviewed journals? (Are they respected, high quality ones?)
- Do the faculty maintain memberships in professional societies? If so, which ones?

Box 3.1. Continued.

- How many students are in the graduate program as a whole?
- How many students are studying topics in fisheries?
- What proportions of the students are in M.S. and Ph.D. programs?
- What courses do you encourage or require your students to take?
- What courses do you teach, and how often do you teach them?
- Does the program offer sufficient courses to fulfill AFS professional certification requirements?
- What is the program's record with regard to placing graduates in fisheries jobs?
- Is tuition payment included in students' financial support?
- Are students provided with health insurance or other benefits?

C. The School

- What other graduate programs or facilities are available at the school that might be useful to students in fisheries?

Questions for Graduate Students

A. Advisor-Graduate Student Relationship

- Are faculty able to spend enough time with their students?
- To what degree do faculty treat students as colleagues?
- What is the reputation of my potential advisor within the program?
- Is my potential advisor actively involved in his or her students' research?
- Is my potential advisor a micro-manager, or more laissez-faire?

B. The Graduate Program

- Are courses in the catalog taught regularly?
- Are the faculty committed to effective teaching, and do they keep their courses up-to-date?
- Which courses would you recommend?
- Do the faculty actively publish work outside of that done by their students? (Does your potential advisor do so?)
- Do the students routinely attend scientific meetings to present their research?

Box 3.1. Continued.

- Do students actively participate in any professional societies? If so, which ones?
- Do the students have a professional or student organization of their own? If so, how active is it?
- Do students usually finish their degrees in a timely manner?
- Is the laboratory and office space available to students adequate?
- What are the demographic characteristics of the students in the program? (Will you fit in?)
- What do students like about the program?
- What do students complain about?

C. The School

- What other graduate programs or facilities at the institution do you find useful?
- How easy is it to find affordable housing in the area?
- Are student stipends adequate to cover living expenses?

career in the long run. Willingness to go anywhere is indicative of maturity and shows your dedication to the profession. Wanting to stay close to home or in a restricted area suggests narrow-mindedness or insecurity; successful graduate students tend to be open-minded and confident. Being flexible in where you live will be a necessity until you start accumulating experience and can start looking for better positions in better places. Exposure to new ecosystems, colleagues, cultures, and issues will expand your understanding and experience. Staying in the same region or worse yet, getting another degree at the same school, will retard your professional growth.

If your interest in a school does have something to do with its location, it may be best not to let that be known. It can be taken to indicate that your interests may not be in research, academics, and dedication to the profession, but rather in fly-fishing, skiing, line-dancing, or whatever else that locale is known for. To be competitive upon graduation, you will not have much time for recreation anyway. Graduate school is a full-time commitment. You may well be better off going to school in an insipid setting where you can get an excellent education with few distractions to divert you. If recreation is your primary focus, graduate school may not be for you. Faculty are also leery of applicants that select a school for personal reasons, such as proximity to one's significant other. Professors prefer students whose sole reason for applying to a program is the educational opportunity it provides; such students tend to be the most dedicated, successful, and productive.

On the other hand, some faculty specifically seek students with regional ties or species-specific affinities, because they believe that such attachments will inspire and focus them. They believe that their passion for a species or topic will enhance the quality of their research and the effort they put into it. These faculty have seen students take an assistantship simply because that was all that was available, but then fail because they were not interested in the topic. Those same students excelled when given a project they had a predilection to. You should discern which type of student you are and find a compatible advisor.

Contacting the Faculty Member

After deciding upon a faculty member with whom you would like to work, or finding an outstanding assistantship vacancy that is compatible with your interests and experience, the next step is to contact your prospective advisor and inform him or her of your availability and suitability for the project or program. It is critical that you make a memorable and favorable first impression (Reese 1999). Therefore, I strongly recommend against initially contacting the professor by telephone. Instead, make your initial contact by email or postal mail, emphasizing your qualifications. The professor can then look at your materials at a time of his or her choosing and can evaluate your suitability for a position. Your packet should include a concise, organized, and detailed letter expressing your interests, career goals, and why you are perfectly suited to the program or advertised assistantship (see Chapter 2 for advice on writing cover letters). Accentuate the ways in which your skills, expertise, and experience match those needed for the research. Give the professor an idea of how motivated and dependable you are by emphasizing your ability to meet or surpass performance expectations in a previous or current assignment that was completed on time and within the limits of the available resources.

Describe your involvement, progress, productivity, and enthusiasm for the work. Remember, the faculty member wants to know if the two of you are going to work productively together with a minimum of problems and inconveniences.

Research potential advisors before writing to them (Reese 1999). Look them up on the Internet, read their recent publications, call their former students, and ask your advisor about them. If applicable, investigate the research topic described in the announcement. Incorporate this knowledge into your cover letter and subsequent conversations. Show that you have done your homework. Make certain that the letter is organized, grammatically correct, and devoid of typographic errors or misspellings. A poorly written letter is indicative of poor communications skills and inattention to detail—it will get you deleted from consideration immediately.

In addition to the letter, include a complete curriculum vitae (CV; see Chapter 2). Include sections on your education, professional and volunteer experience, internships, professional society affiliations and service, short courses, skills, certifications, honors and awards, publications, and presentations. Also append photocopies of your GRE scores and university transcripts (which should include a final or current grade point average; if not, calculate and provide it), plus the names, titles, and contact information for three professional references who can describe your academic and professional abilities as well as your character (professors in particular, but also supervisors or colleagues, and not clergy, politicians, or friends). At this stage, you do not yet need written letters of recommendation. Delay asking for these until submission of the formal application, at which time you will know exactly what they need to address; some schools ask specific questions or use a form. You do not want to hassle your references for multiple versions. Enclose an example of your technical writing (e.g., senior or M.S. thesis, term paper, or reprint) in your packet to illuminate your communications skills. Advisors spend countless hours teaching graduate students how to write; showing that you can already write well will significantly enhance your prospects. Also include any other items requested in the assistantship announcement. Email, overnight, or priority mail this pre-application packet to the faculty member.

Telephoning the Faculty Member

If you have the necessary qualifications and have presented them favorably, the faculty member will soon be calling you. If not, give him or her at least a week or two to read your material before making follow-up contact. When you do call, first try to set up an appointment for a subsequent call; the professor may be busy at the moment you call and may prefer to talk to you later. Better yet, schedule the call by email. Call promptly at the appointed time, not early and most certainly not late. Address the professor as "Dr." until told otherwise. Start the conversation by reiterating your background, skills, expertise, and experience, and how these match those needed to do the research. Do not presume that the professor has studied your pre-application in great detail and knows who you are. Answer any questions forthrightly and honestly. Organize your thoughts and speak clearly. Think of this conversation as a phone interview (see Chapters 4 and 5).

Do not start by asking the professor to describe the project (if responding to an assistantship announcement). It might seem like a convenient way to start the conversation, but professors dislike having to describe a project for the umpteenth time. If you have prepared appropriately, you should already have a good idea of what the study involves. Remember, *you* are applying for the assistantship and the professor will have numerous candidates to choose from. If you come across as uninformed, you will not be viewed as favorably as an applicant who already has a grasp of the project and recognizes the relevance of the research. However, you should have some *specific* questions about the study (and university) ready in case the professor asks if you have any (Box 3.1). Make sure that they are insightful to show that you have done your homework, not just things you should have already looked up on the Internet.

If this initial conversation goes well, enlist your references, particularly your advisor, to contact the professor and speak on your behalf, especially if they already have a pre-existing relationship. Faculty are much more comfortable accepting a student who has been endorsed by someone they know and trust than accepting a student who looks great on paper, but about whom they do not have a personal guarantee. Exploit this by using your advisor's connections. If you have any other direct connections, such as a coworker, supervisor, or fellow student who knows the professor, be sure to use these as well. Personal connections can be crucial for getting into graduate school.

Visiting Your Prospective Advisor

Schedule a visit if it seems that the professor is interested in you and you are still interested in the position (Fischer and King 1998). A professor is much more likely to extend an offer to someone that they have met and who appears to be someone that they can get along with and work productively with, than a prospect that they know only through correspondence. A visit is also an opportunity for you to assess the school, its resources, the project, and the faculty member (Allen 1993). Be sure to talk to the professor's current students (Box 3.1); they can offer the best insight into what your life will be like if you go to graduate school there. Of course, you have to make a good impression during your visit. That is, you need to come across as mature, intelligent, knowledgeable, dedicated, productive, and easy to work with. See the interviewing tips described in Chapters 4 and 5; most of them apply to this situation as well. Dress nicely and pay attention to personal hygiene.

The Decision

If you are qualified and have done well in your interactions with your prospective advisor, you have a good chance of getting an offer. If you do, give it careful consideration, weigh your options, and make a decision within a few days. Call the faculty member and let him or her know what you have decided. If you decide to pass, let the faculty member know promptly, so that he or she can offer the assistantship to someone else. Be courteous and considerate because the fisheries profession is small and you likely will deal with this individual again in the future. If you decide to accept, submit the official application (if you have not already been required to do so), schedule a starting date, make plans to relocate, and get ready to begin one of the most exciting, challenging, and satisfying periods of your life.

Dealing with Deficiencies

All of the above is based on the assumption that you have been a stellar undergraduate student and therefore have excellent grades and GRE scores, as well as loads of experience, professional service, honors, awards, and outstanding references. But, what if you do not? For example, perhaps you partied a bit too much as a freshman, flunked a few courses, and your GPA was never able to fully recover. Or, maybe you are simply not good at standardized tests and your GREs reflect that. Having a deficiency in one category is not usually a deal-breaker if your other categories compensate for it. For example, a low GPA, especially one that shows consistent improvement after a poor freshman year, can be offset by high GREs (and vice-versa). Experience, good references, honors, awards, publications, presentations, and professional activities are all important factors, and depending on the project and perspective of the advisor, may help bail you out. Therefore, be sure to highlight your strengths.

Having several deficiencies will not necessarily exclude you from getting into grad school either. It will, however, likely limit you to less desirable opportunities such as those at less prestigious schools, teaching assistantships, or research assistantships that entail work unrelated to your research topic. These can be golden opportunities for you to show your true potential and still make fisheries your career despite past mistakes and deficiencies. If you excel, you might have your assistantship converted to a project-related research assistantship in your second year. Moreover, teaching experience and familiarity with another research topic will pay off in the long run.

Another option is to be admitted as a nondegree graduate student unsupported by an assistantship and enroll in several graduate courses. This can be costly compared to getting an assistantship, but you can thereby get to know the faculty, participate in professional activities, volunteer on other students' projects, ace the courses, and impress the heck out of everybody such that when the next assistantship becomes available faculty will select you despite your deficiencies. A variant of this strategy is to work as a technician for an agency and impress your supervisors to such an extent that they fund a graduate research project tailored to your experience and knowledge. The agency benefits by having a known individual conduct the research and by subsequently gaining a more educated and promotable employee. This strategy is something of a long shot, but occurs more frequently than you might expect. The bottom line is that deficiencies are surmountable if you are dedicated, hard-working, and willing to make sacrifices. A career in fisheries is well worth it.

Sources of Additional Information

Most announcements for graduate student positions are posted to the AFS Job Center Online (under *Student Opportunities* at http://www.fisheries.org/html/jobs.shtml) or the Job Board hosted by the Department of Wildlife and Fisheries Sciences at Texas A&M University (under *Graduate Assistantships* at http://www.wfsc.tamu.edu/jobboard). You can also broaden your search by asking your advisor or colleagues to watch for announcements and forward them to you, or by signing up for email listserves that routinely receive announcements. Some useful listserves include ECOLOG (https://listserv.umd.edu/archives/ecolog-l.html; sponsored by the Ecological Society of America) and those listed on the AFS Computer User Section web site (http://www.fisheries.org/cus/cuslistsnew.htm; especially AFS-L). The Education Section of AFS maintains a web site focused on the educational requirements of fisheries professionals (http://www.fisheries.org/education/programs.htm). This web site lists colleges and universities that have graduate programs in fisheries and related fields and is useful for researching potential schools.

References

Allen, M. 1993. Guide to choosing a graduate school. Fisheries 18(2): 30–31.

Fischer, R. A., and S. L. King. 1998. Suggestions for new and aspiring graduate students in wildlife science. Wildlife Society Bulletin 26: 41–50.

Gibbons, J. W. 1998. Graduate education at a field research laboratory: facing the challenge. Herpetologica 54(supplement): S21–S30.

Grossman, G. D. 1998. Notes from the blackboard: choosing the right graduate school and getting the job you've always wanted. Fisheries 23(9): 16–17.

Orth, D. J., and I. R. Adelman. 1996. Finding a graduate program in fisheries. Pages 19–33 *in* T. D. Hill and R. M. Neumann, editors. An AFS guide to fisheries employment. American Fisheries Society, Bethesda, Maryland.

Reese, K. P. 1999. What aspiring graduate students should not do. Wildlife Society Bulletin 27: 254–255.

Rossman, M. H. 1995. Negotiating graduate school: a guide for graduate students. Sage Publications, Thousand Oaks, California.

Zale, A. V., R. L. Simmonds, Jr., and R. T. Eades. 2000. Getting a job or assistantship: how to surpass the competition. Fisheries 25(6): 24–30.

Chapter 4

Fisheries Employment in State Agencies

RICHARD T. EADES

Every state has an agency with a department responsible for its fisheries resources, so it should be no surprise that many fisheries professionals work for a state agency at some point in their careers. Most states have just one agency to oversee their fisheries, but some coastal states have separate agencies for freshwater and marine resources (e.g., Virginia). Whereas every state has a department or division responsible for fisheries management, the name of the parent agency varies from state to state, so finding the name of a potential employing agency might be a little tricky. Some states have a Fisheries Commission or a Game and Fish Department, and finding these agencies is rather straightforward. However, in other states the fisheries entity may be embedded within a department for which the connection to fisheries is less clear (e.g., the Texas Parks and Wildlife Department). In any case, each state provides a full list of its agencies on its official government web site.

The variation among state agencies is usually the result of politics; often the election of a new governor leads to organizational restructuring or combining of agencies in an attempt to reduce costs, redundancy, or bureaucratic red tape. Many states started with an independent fisheries agency, but over time decided it was best to combine it with other departments responsible for things such as wildlife, nongame species, habitat protection, parks, water quality, or boating. Each natural resources agency has gone through so many stages of growth and restructuring that no two are alike. Nonetheless, they are still similar enough in their mission to manage aquatic resources that a good education and some practical experience will qualify someone for an entry-level position in most of them.

Considerable variation also exists in the composition of state fisheries departments or divisions. Obviously, the local resources dictate much of their focus and structure. For example, coastal states have sections and personnel dedicated to marine resources, and states with separate marine resource agencies may have sections for recreational and commercial fishing, estuarine and ocean resources, or shellfish and finfish. Similarly, states along the Great Lakes have staff dedicated specifically to managing those waters, and some states have programs focused on large river systems that fall under their jurisdiction (e.g., the Mississippi and Colorado Rivers). In addition, hatchery programs are often a major element of state fisheries departments and may include both sport and nongame species for coldwater, warmwater, or marine environments.

Working for a state agency has both advantages and disadvantages in comparison to federal, academic, and private positions (see Chapters 5, 6, 11, and 12). Permanent state agency employees enjoy greater job security than employees in the private sector and usually receive generous benefits packages. However, state agency employees may receive lower salaries than individuals employed by the federal government or private entities. State agencies have fairly stable funding sources, so employees are under less pressure to obtain grants and outside funding than academic and private sector employees. State fisheries employees are often encouraged to publish peer-reviewed articles, but face much less pressure to do so than individuals at academic institutions. Finally, employees in state agencies typically have more opportunities to transfer to different job locations than professionals in the private sector, but fewer chances than federal employees.

Types of Positions Available

Fisheries departments or divisions are usually separated into sections, such as management, production, and research, and each section includes various types of jobs with different responsibilities. A 2001 survey of inland state agencies indicated that, on average, 36% of their fisheries personnel worked in management positions, 32% worked in fish production, 7% were in research positions, 5% were administrators, and the remaining staff were involved in a variety of other areas such as nongame species, habitat, and education (Gabelhouse 2005). Because of the differences in duties and required skills, a majority of professionals in state agencies follow a career path through either the management or production sections, with few cross-overs. The same is generally true for freshwater versus marine resources; most professionals focus on one or the other for much of their career.

As the name implies, fisheries management staff are primarily responsible for the management of fisheries resources, but are equally concerned with managing people—the users of those resources. Managers strive to understand the ways in which people interact with aquatic resources and how to manage the use of those resources to provide the greatest benefits to both people and the environment. For example, managers often collect and analyze data to determine appropriate regulations for species targeted by commercial and recreational fishermen (e.g., size limits, creel limits, closed seasons, gear restrictions). Entry-level positions in fisheries management include field technicians that are primarily responsible for collecting data, survey clerks that interview anglers or other resource users, and lab assistants that prepare fish scales, spines, and otoliths for age and growth analysis. These positions do not typically involve report writing, detailed data analysis, or decision making, and supervisory responsibilities will probably be limited to overseeing the work of temporary employees. Advanced positions, such as biologists, program managers, and district supervisors, will have increasing responsibilities for data analysis, report writing, supervision, budgeting, and making recommendations for management actions.

Staff at hatcheries and fish production facilities are responsible for raising aquatic organisms for release into the wild. Depending on the location of a facility, production managers may oversee the propagation of a wide variety of freshwater and saltwater organisms. Whereas inland states typically raise game fish species for stocking into recreational fisheries, coastal states may raise shellfish and finfish to support recreational as well as commercial fisheries. Some states may raise threatened or endangered species for conservation purposes, but these efforts are more common at federal facilities. Duties of entry-level personnel in fish production may include monitoring water quality; feeding, sorting, grading, transporting, and stocking fish; and collecting broodstock for spawning. Advanced positions, such as biologists and hatchery managers, will have greater responsibilities in supervision, budgeting, and planning for various production needs.

Research is defined as investigation or experimentation aimed at the discovery and interpretation of facts, revision of accepted theories or laws in light of new facts, or practical application of theories. Fisheries research professionals in state agencies seek to answer questions that can help address resource problems. At the simplest level, a fisheries researcher tries to answer the question, "How can we do things better?" Managers use research findings to assist them in making sound decisions about the resources. For example, researchers may determine the most cost-effective methods for raising fish or the optimal sizes and numbers of fish for stocking programs. Researchers often spend a lot of time in the field studying a particular issue, but usually spend an equal amount of time in the office examining historical data, the scientific literature, and computer models. Advanced research positions, such as section leaders, have responsibilities for writing grant proposals, planning, budgeting, supervising projects, and publishing. Similar to other sections of the agency, the top person may have primarily administrative duties and be classified accordingly (e.g., assistant chief).

A state agency cannot function without capable administrators to develop and monitor budgets, approve purchases, determine personnel needs, and make final decisions on matters of importance (see Chapter 13). Administrators are usually seasoned fisheries professionals with a broad range of experience and training. They consider the recommendations from their assistants and other staff in deciding on the ultimate course of action to be followed. The top administrator (e.g., division chief) is usually responsible for presenting recommendations to the agency's director and commission for approval.

Every state agency is different, and in each agency there may be a variety of specialized positions outside of the management, production, and research sections. Common examples include positions devoted to aquatic vegetation control, nongame species conservation, aquatic habitat enhancement and restoration, and aquatic resource education. Education positions are particularly common because agencies recognize the importance of an educated public, and thus seek to create programs, particularly for children, that demonstrate the values of healthy aquatic environments. Positions in aquatic education may be housed within the fisheries division or be part of a separate division dedicated solely to education.

Depending on the state's governmental structure, additional positions focused on aquatic resources may be available through other state agencies or other sections within a larger department, such as a Department of Natural Resources or a Department of Environmental Protection. Such positions are often concerned with monitoring water quality, point source pollution discharges, water use, or other activities that affect aquatic resources. Some positions may involve sampling fish assemblages and collecting fish tissue for contaminant analysis. A state's Department of Health may also employ individuals with fisheries backgrounds to collect and analyze fish tissue samples and determine if fish consumption advisories are needed. Agencies with responsibility for a state's parks, such as a Department of Parks and Recreation, may also employ fisheries professionals as naturalists to provide educational programs to park visitors.

Starting Out

If you are looking toward a career in state agency fisheries employment, you should think about what your main interests are and how they relate to the types of positions available in state agencies. Try to decide whether you like working in hatcheries or whether you are primarily interested in management of sport fisheries, habitat, or nongame and endangered species, and to what degree you want to be involved in research. Whereas many people will tell you that your education and experience should be as broad or well-rounded as possible to make you a good professional, you should have some idea as you begin your career as to what path you wish to follow. Of course, you may have to be flexible at first to "get your foot in the door," and you can certainly change paths in the future as your skills develop and your interests change. If your interests revolve around being outdoors, traveling, and collecting fish, apply for field jobs, not ones that might have you sitting in an office going through data for weeks at a time. Also bear in mind that in the marine environment, the boats are bigger, the water is deeper, and the environment is often less friendly. If you are prone to seasickness or would rather not go several days without standing on solid ground, be careful which coastal positions you choose.

One way to test the waters of different types of jobs, or a route to employment when permanent positions are hard to find, is contract positions with state agencies. These positions are temporary and typically focused on specific issues. Contract positions may last from a few months to several years, and some may have recurring potential for extensions, but they may or may not provide benefits such as health insurance and paid leave. These positions allow you to gain experience, acquire new skills, make professional contacts, and earn money while you search for a permanent position.

As you start your job search, you need to know some of your personal preferences. For instance, are you willing to work anywhere in the United States? If so, there are fifty states with potential employers. If you love trout fishing in the mountains and do not want to work anywhere that does not provide that opportunity, you have fewer employment possibilities. If you are interested in marine resources, do you prefer the East Coast or the West Coast? Be sure to research locations before you start applying for jobs so that you are sure you are applying for jobs in places you are willing to live for at least a couple of years. It might take time to find another job if the location of the first one is not a good fit for you.

When considering job locations, you should also think about the various fish species you have worked with in graduate school or previous work experiences. That can have a big influence on your competitiveness for a job. Depending on the state, management emphasis may be directed toward some game or commercial species more than others. If your interests and expertise are in walleye *Sander vitreus* management, there will be more job opportunities in the Midwest. If you prefer to work with salmon, try the Pacific Northwest or New England. If you are more interested in managing farm ponds for trophy largemouth bass *Micropterus salmoides* and bluegill *Lepomis macrochirus* or if you like charismatic marine species like tarpon *Megalops atlanticus*, you will likely be happier and more successful job hunting in the Southeast.

All that being said, it is helpful in your job hunt to have some flexibility. If you are too focused on working on a specific species or in a specific location, you may be hard-pressed to get a full-time permanent position with a state agency. Whereas it is fine to have a goal to someday work in a specific position in a specific place, if the person who currently has that position has no plans for promotion, moving, or retirement in the next 15 or 20 years, it will be a long wait to have a shot at that dream job! Perhaps you can find a similar job across the river or down the coast in another state. Maybe you can still live in that location, but work on something different. The key is to make yourself competitive for jobs you would enjoy, in places you would like to live (see *Keys to Success*, Chapter 15).

Acquiring Needed Qualifications

Your first objective is to acquire the qualifications needed for an entry-level position and to be competitive enough to get an interview (Zale et al. 2000). Experience is paramount, so do whatever it takes to get as much as possible before, during, and after getting your degree (see also Chapter 1). Volunteer experience is perfectly acceptable if no paying positions are available and internships can be the most valuable volunteer experiences.

Coursework requirements for most fisheries positions in state agencies are comparable to those required for professional certification through the American Fisheries Society (AFS; see Chapter 15). In addition, certification can provide bonus application points or promotion credits with some agencies (Pegg et al. 1999). The greater the diversity of courses you have taken, the better you may do in an interview, because you never know which pieces of information will help you answer a question. Furthermore, statistics courses are invaluable because established agency personnel recognize the need for such expertise, but often do not have enough of it on staff and look for new employees to provide it.

Publications on your resume or CV are a strong indication that you are smart, dedicated, hardworking, and follow through on projects. Publishing your thesis research as soon as possible is critical considering the competitiveness of the job market. Whereas some agencies may not consider publishing by their current personnel to be a high priority, supervisors nevertheless recognize it as an effective means of gauging applicants. Publishing also includes popular articles, and agencies repeatedly mention their value because magazine articles and press releases are important tools for managers. You can easily write a few popular articles about your current research for the local newspaper or your state's outdoors magazine. Writing skills are so important that some agencies now include a written essay question as part of the job interview. Giving the right answer is important, but not nearly as important as showing that you can write.

Presentations are evidence of your public speaking abilities, which are paramount in state agency jobs. The ability to perform well in a public hearing may make more difference in getting a management action implemented than the biological validity of that action. Agencies want people who can communicate with the public as well as with other biologists. Make as many presentations as you possibly can. Possible venues include AFS meetings at the chapter, division, section, or parent society level; departmental seminars; state academy of sciences meetings; other professional and scientific society meetings; and special topics symposia. Many of these occur within driving distance of your current location each year. A few agencies now require a presentation as part of an interview, most commonly for research positions.

Involvement in professional societies does not accrue application points in most states, but some employers look favorably upon such involvement as a sign of professionalism (see Chapter 15). It also indicates that the applicant is dedicated to a career in fisheries rather than just a job. For example, a person who has served as an officer in an AFS unit has shown a willingness to take on added responsibilities and accept leadership. Networking through involvement in professional societies is critical to your career, but not as important as it once was for getting interviews. Because states have objective hiring guidelines, it is no longer whom you know but what you know that counts in making the first cut. Knowing somebody may help you get the job over comparably qualified candidates not known by the hiring panel, but you still need to be selected for an interview first.

Finding Job Vacancy Announcements

Job vacancy information is now readily available on the Internet. Twenty years ago, applicants had to make a lot of phone calls, write letters, check bulletin boards, and pay job service companies in hopes of finding out about job openings. Today, a little time on the Internet can provide just about any information you need on job openings, state agencies, and the particulars about a state, such as the cost of living in different communities, housing, climate, fishing and hunting opportunities, and anything else that might help you decide if a job is right for you.

Despite the efficacy of the Internet for finding out about openings, you should not rely on it exclusively. Many state job openings are posted at the AFS Job Center Online and other similar web sites, but not all. Most Coop units (see Chapter 7) and university fish and wildlife programs are on agency mailing lists and either post vacancy announcements on a jobs board or keep them in a special file. If you have access to these, check them at least weekly; some positions may be advertised only on paper and not on the Internet. In addition, you should call the personnel office of each agency you are interested in directly to find out the best way to keep track of openings and application procedures. Some states may not post all jobs on the Internet, and others may be tardy about keeping their web pages current. Your professional connections are also important in your job search. Although they may not help you get an interview, they can help you find out about upcoming vacancies before your competition does. This edge can give you extra time to tailor your application to that particular job and to become more knowledgeable about the position and how best to interview for it.

Some states do not advertise vacancies, but select candidates from an internal register instead. For example, an agency may use the applicant pool from a previously advertised position to fill future similar vacancies, or they may open their internal register for a period of time to new applicants and then close it for a year or more. Candidates must typically pass an exam to get on the register, and once on the register they are notified of job openings and can compete for them. Registered candidates must maintain their status by notifying the personnel office at prescribed intervals (typically annually) and should be sure to update the resume or CV they have on file as needed. Some states openly advertise positions but require that candidates be on their register to apply, and getting on it may be impossible before the vacancy closes. Other agencies openly advertise positions to any qualified candidate. You should request application materials and information about application procedures from agencies that you are particularly interested in before positions open. That way, when a vacancy is announced you will be prepared and will not have to wait for the forms and run the risk of missing a tight deadline. However, do not complete the application until you see the announcement, as you will want to tailor it to the specific position description.

Tailoring Your Application

Some state agency vacancies receive up to 200 applications, but at most only a handful of applicants are invited to interview. The job application is the key to getting an interview. Applications are often scored objectively by someone in the personnel office, rather than by fisheries staff, and they use the position descriptions in the announcements for scoring criteria. Therefore, you need to tailor your application and resume or CV to that specific job description (see also Chapter 2). Avoid copying the same boilerplate you might have been using for other applications. To make it obvious that your qualifications match those in the position description, describe your qualifications using the exact phrasing in the description. This strategy applies even when the people rating the applications know you and your qualifications. An application is an objective screening instrument that must be interpreted strictly and impartially to pass legal muster. Far too many qualified candidates exclude themselves from consideration for an interview by not following this simple guidance. For example, a district fisheries biologist with more than 15 years of experience was disqualified from a similar position in another state because he did not specify on his application that his current position was full time. Never assume that the personnel office will deduce anything beyond what you state explicitly (Zale et al. 2000).

Make sure that your application, transcripts, and resume or CV specify explicitly the emphasis in your education. When scoring your application for a fisheries position, the personnel office or hiring supervisor will likely give you a higher score for a degree in fisheries or fisheries and wildlife than one in biology or forestry. In addition to the application materials you send to the personnel office, send a copy to the hiring supervisor with an appropriate cover letter.

Some state applications require you to submit a written summary of knowledge, skills, and abilities (KSAs) or knowledge, abilities, skills, and other considerations (KASOCs). These summaries ask you to describe precisely how you qualify for the position, usually in addition to the application and your resume or CV. Related questions may also be included in the job announcement. For example, you may be asked how your KSAs will enable you to complete specific job responsibilities. Answer each question completely and accurately, without lying or exaggerating. If a state has a special qualifications section on its application, pay particular attention to it.

Emphasize experience that specifically corresponds to that described in the position announcement. For example, if you are applying for a management job, emphasize your experience related to planning, budgeting, supervising, data analysis, and report writing, not technician duties. Technician level skills, such as fish sampling and boat maintenance, should still be included but should not be the primary focus. If you lack the experience required by positions you are interested in, seek opportunities for such activities in your present position.

Most states ask for a list of references (typically three), which should be carefully selected individuals who can fully discuss your education, experience, personality, and work habits. The reference list is usually comprised of past professors and supervisors, not your parents or best friend. If you cannot name three such people who would tell a potential employer that you are knowledgeable, skilled, hardworking, and that they would be remiss to not hire you, you have a problem that needs to be fixed in a hurry. Have your list prepared, make sure the contact information is accurate, and ask the people for their permission to be included. Make it easy for the potential employer to contact your references.

Reference checks are more important than you might think. Some states are required to check references, even if they know you and your qualifications. If an employer is faced with a tough choice between two applicants, the references can be the deciding factor in who gets the job. If a former employer of one applicant says that the person was often late for work and had a hard time getting along with coworkers, it is very likely that the other applicant will be offered the job. Make sure your references know your qualifications and provide each of them with a copy of your resume or CV.

If you are not selected for an interview, call the hiring supervisor to ascertain why, but do not expect him or her to remember your name or pull your application. Ask what qualifications the interviewees possessed so that you can identify your deficiencies. If you think some of your qualifications were overlooked, rewrite your resume or CV to better emphasize them for the next opportunity.

Preparing for the Interview

If an agency decides to interview you, the hiring supervisor will either send you a letter providing the date and time of your interview or call you to set an interview time. If called, you have an opportunity to make a good impression. You should be happy that the supervisor is interested in you and taking time to personally call to invite you for an interview. Be prepared, be excited, say thank you, and be agreeable. Most agencies are accommodating and will set an interview time that works for you, within reason. Typically, applicants traveling the farthest to interview are given first preference on interview times to fit flight schedules. When you apply for a job you should know how you would get to the interview location. Keep in mind that it could be a month or two between the time you submit your application and when the interviews are held. You can be sure the interviews will take place between 8:00 a.m. and 5:00 p.m. on a weekday; the interview panel will not come in on a Saturday so that you can avoid missing classes or work. When the supervisor tells you the interview date and asks what time you would prefer, have an answer. You will make a good impression by saying, "Thank you. I'd prefer a morning interview. I'll fly in on Monday evening and return home Tuesday afternoon." You will make a bad impression by saying, "I'm not sure. I'll check my schedule and figure out how I'm going to get there. I'm not sure I can get off work. If I decide I'm still interested, can I call you back tomorrow with a time that works for me?" That tells the potential supervisor that you do not plan ahead, that you are indecisive, and that you are not too excited about the opportunity.

Some supervisors will conduct phone interviews, but the practice is rather rare. Some may conduct phone interviews as a first step, and then invite the top applicants for a follow-up meeting. Supervisors want to meet applicants before making a hiring decision, and you should want to meet your potential supervisor and coworkers and see where you could be working. If you are offered a phone interview, it is fine to accept, but in general it is best that you do not request one. Asking for a phone interview might send the message that the interview is not worth much time, effort, and expense to you.

After you have agreed to an interview, learn everything you can about the position. It is likely that many interview questions will be based on responsibilities of the position and current issues. You should talk to someone familiar with the agency and area, such as current and past employees, the hiring supervisor, or others who recently interviewed there. This is where your professional connections often pay off. You can also read pertinent reports and publications from that office, read the local newspaper, and review back issues of the agency's outdoors magazine. Study topics that your inquiries and investigations identify and you will impress the interview panel, not only with your knowledge, but also your preparedness. If you lack any desired skills or experience, try to find a training course, volunteer to help with appropriate field work, or read up on the topic prior to your interview.

The Interview

Whereas there are no universal procedures for interviewing candidates, this section describes the most common format. Individual supervisors may adapt procedures within the limits of state personnel policies to fit their personal preferences. However, federal law prevents any interviewer from asking any questions of a personal nature, such as a person's religious beliefs, marital status, political affiliation, or race.

When you arrive for an interview, be pleasant to administrative and support staff; they are potential future coworkers. The last person you want as an enemy is the office receptionist or administrative assistant. Furthermore, although they are not part of the interview panel, they may be asked informally for their opinions of the candidates.

Your interview will probably take place in a conference room, most likely at the office where the position is located or at the agency headquarters. The hiring supervisor typically leads the interview, and three or more panelists will ask you questions. Panelists are typically supervisors from other offices or divisions. Some states require one panel member from another agency division, such as wildlife or education, to provide an outside perspective. These panelists are not trained fisheries professionals and may not know the correct answers to some technical interview questions. Their role is to evaluate your personality and presentation and whether you are someone they think would work comfortably and productively with others. This is especially true for positions in regional or headquarters offices, where fisheries personnel comprise only a portion of the staff. Dress appropriately, be polite, attentive, and congenial, and answer the panelists' questions directly and completely. Sell yourself and be confident, but do not exaggerate or come across as arrogant.

One especially important thing the panel will be looking for from you is enthusiasm and you need to be sure to give the impression that you want the job. You should let the panel know not just that you are the right person for the job, but that the job is right for you. An attitude of excitement and "When can I start?" will be viewed much more favorably than an attitude of "I really don't care if I get this job because I've applied for a bunch of other ones too."

The hiring supervisor usually starts by going over the ground rules. You should realize that the panel might be on a strict time schedule to interview everyone, so pay attention to the number of questions and the amount of time allotted. If there will be ten questions and you have one hour, plan accordingly. That does not mean you should spend precisely six minutes on each answer, as some questions require longer answers than others. If you are asked to define proportional stock density, answer that in a minute or two. If you are asked to describe the education and experience you have that make you the ideal person for the job, spend ten minutes on it if you have many relevant things to say. You may want to ask the panelists questions, but make sure you answer all of theirs first. You can always call back later and ask your questions, but you will not be given a second opportunity to answer theirs.

While you are answering the questions, the panelists will take notes and perhaps grade your responses. The exchange does not resemble a conversation, as you may be the only one talking for 5 or 10 minutes. Panelists may ask follow-up questions or want you to elaborate on or clarify something. Some panelists will nod and smile at your answers and make you feel like you are doing well, whereas others may not give you any feedback at all. Do not panic or allow yourself to feel intimidated.

When asked a question, try to start answering after only a few seconds of thought, structuring your response in an ordered and logical sequence as it unfolds. You should use clear language and avoid affectations and idiosyncrasies. Biologists are asked tough questions every day and their reputation depends on thinking fast and giving good answers. Do not try to write out the interview questions and formulate responses on paper. Whereas the answers derived in that manner may be excellent, the long pause will give the panel plenty of time to stare at the walls and wonder if you will do the same thing every time someone asks you a tough question. If you do not know the answer to a question, say so and move on. The ability to admit that you do not know something is a good trait, and by no means should you pretend that you are knowledgeable about something that you are not.

Demonstrating that you are ready for the rigors of the job will go a long way with the panel. Answer questions with real-world, practical answers. You will probably never hear a state administrator say that his or her fisheries department is overstaffed and underworked. The unwritten motto of most departments is "Do more with less," and if you understand that and have experienced that, you will do better in interviews. For example, in an interview for a management biologist position, applicants were told about a perceived problem at a small lake and asked what they would do to develop a recommendation for their supervisor so that he or she could respond to anglers' questions. Some applicants appropriately answered that they could have a recommendation within 24 hours, after a quick electrofishing sample and a review of previous data. Others described elaborate multiyear research projects that would result in a solid recommendation in 2 or 3 years, which would be unacceptable in this situation. Hatchery workers are faced with problems requiring similarly fast responses and interviews for hatchery positions can involve some of the most applied questions. For example, you should be prepared to answer questions such as:

1. *How do you recognize and treat bacterial kidney disease?*

2. *What would you do if the fresh water supply was cut off or contaminated?*

3.   *What if cormorants and otters were eating the fish?*

4.   *Can you calculate how much chemical you would need to add to a raceway 50 feet long, 20 feet wide, and 5 feet deep to achieve a 2-ppm concentration?*

You should also keep in mind that all agencies have a chain of command that is followed. Whereas you want to impress the panel with your initiative and willingness to tackle tough problems, you need to show proper deference and acknowledge that supervisors make many of the substantive decisions. If you are asked how you would handle a problem or design a new study, include in your response that you would ask others, especially your supervisor, for input.

There are three types of questions that you should expect in an interview. The first type is aimed at finding out about your "people skills." You might be asked how you would deal with a problem employee, a personality conflict with a coworker, or a troublesome constituent group. Secondly, you might be asked to describe a real-world management problem that you have been confronted with and how you solved it. Finally, you may be asked to describe your strengths and weaknesses, either as you see them or how you perceive others see them.

As in your application, never assume the interview panel knows everything you have done. Many applicants, when sitting in front of an interview panel of coworkers, make the mistake of starting off answers with "You guys know me, and know that I've spent a lot of time...." Maybe they do, but maybe they do not; it is not worth taking that chance. It may seem strange to describe your experience to coworkers, but the panel is comparing your answers to those of other candidates. If another candidate spends 10 minutes detailing an amazing record of accomplishments, and your answer is simply, "You guys know what I've done," do not expect to score very well on that response.

Another problem is answering with "we." Remember that the interview panel is interviewing *you*. They want to know what you have done and how you would handle a situation. Some interviewees answer, "Where I work now, we do..." or, "When I worked in Texas, we used to...." The problem with these responses is that the applicants are not telling what role they played and it often sounds like they are taking credit for other people's work. If you are asked to describe your experiences with public hearings, do not say, "When I was in Georgia, we went to public hearings several times a year...." Specifically describe what you did at the hearings (e.g., presided over the meeting, gave a presentation).

The advice suggested here is not unique to fisheries. Surveys have shown that employers from a variety of fields consistently rank relevant experience, professionalism during the interview, fit with the company culture, education, and enthusiasm as the top attributes of successful job candidates. Furthermore, a candidate's ability to relate his or her past experience to the job being applied for is always an important factor in the hiring decision.

After the Interview

After your interview, hopefully you will receive a job offer. If you do, realize that the agency has completed the evaluation process and they expect you to accept their offer. If you turn their offer down, that means more work for the panel and they may not remember you favorably in the future. Of course, you do not have to accept a job offer, but if you know at some point, for whatever reason, that you will not accept, decline the interview and allow somebody else the opportunity. Likewise, if you decide after the interview that you would not accept the position if offered, let the supervisor know before an offer is made.

One of the big stumbling blocks in the hiring process, especially for advanced positions, can be the starting salary. Most job announcements show the salary range for the advertised position. For example, the announcement may indicate that the annual salary range is $28,000–$42,000. This usually means that employees in the position receive a starting salary of $28,000, and over time they can earn $42,000 with pay raises. It does not mean that they will pay you $40,000 if you say that is how much you deserve. Departmental policies and state laws control salaries, and many states explicitly explain that employees new to the agency must start at the minimum salary listed for the position. If that is not acceptable, do not apply for the position! The job market is competitive and supervisors know that there are applicants who would gladly start at $28,000 in order to get a permanent, full-time position with benefits. However, note that if you currently have a similar job in another state, many states will at least match your current salary, and may offer a predetermined percentage increase.

Persevere

If you find that you do not have enough experience to be competitive for a permanent position, look for positions that provide more experience, professional contacts, and knowledge, including low-paying, temporary, or volunteer positions. Any position could provide you with an opportunity to learn something new that may help you answer that one interview question that lands you the next job. Most state agencies hire temporary employees, especially during the summer, to conduct creel surveys and assist with field work. These are great opportunities for students to gain practical experience while still working on their degrees. For graduates, taking a temporary position may not sound attractive, but it is certainly better to be gaining relevant experience than to be unemployed or working in an unrelated field.

Even if you are qualified, you may apply for dozens of permanent jobs and go to four or five interviews before being hired. Do not get discouraged—the more you interview, the better you will be at it. If you interview but are not selected for a position, follow up with the hiring supervisor to find out where you fell short and then reapply the next time the agency has a vacancy. Remember that more than one supervisor served on your interview panel. One of them may want you for another vacancy in their region, so try to impress everyone during the interview. If you are highly qualified, you will be remembered the next time there is an opening. When you get a second chance, you have an edge over the competition, so take full advantage of it. You will know the interview protocol and what kinds of questions to expect, and some panelists may already know you. You can relax, do well, and land the job. The bottom line is that hard work and preparation pay off in a competitive environment. Acquire the needed credentials, find appropriate opportunities, assert a positive attitude, and prepare fully for each opportunity you get.

Advancing to Higher Positions

If the competition for an entry-level position sounds tough, just wait until you look to advance to higher positions. The higher up in a state agency you go, the fewer opportunities there are. Starting from the top, each state has only one person in charge of the fisheries division (a state with a separate marine resources agency may have two such persons). If you want to be a freshwater fisheries chief, there are only 50 opportunities. Most state chiefs have a few assistant chiefs as well. Assistant chiefs supervise one of the division sections, such as management, production, or research. A variety of program managers, hatchery superintendents, regional managers, and district supervisors report to the assistant chiefs. Below them are biologists, technicians, and hatchery workers. In a state fisheries department, there may be 100 entry-level workers who report to 10 hatchery managers and district supervisors, who in turn report to two assistant chiefs, who then report to the chief. Obviously, getting one of those first 100 jobs is much easier than advancing to one of the 10 supervisory positions or three administrative positions.

Again, flexibility is important in your job search. With limited advancement opportunities within a state, you can increase your opportunities by being willing to relocate. However, remember that you will be most competitive for and most likely to obtain advanced positions in a given specialty if you are already working in that specialty (e.g., management, production).

Apply the same techniques for landing advanced jobs that you did in landing an entry-level one. Try to get as much experience as possible and do things that make you stand out from the competition, such as publishing papers, giving presentations at conferences, and perhaps serving as an officer in a professional society. Advanced degrees are increasingly important as a highly preferred or required qualification for advanced positions in state agencies, so acquire a graduate degree if you do not already have one.

Too many professionals think that they will get a promotion just by being somewhere for a long time. For example, most management sections are broken down into geographic districts with a supervisor and assistants. When a supervisor is promoted to an administrative position or retires, too often the assistant thinks he or she will be promoted because they have been a district assistant for 10 or 20 years. Although this does happen, it is never a given. If you do not continue to learn, grow, and improve throughout your career, you will be outcompeted for promotions by colleagues from other districts or states.

One of the big challenges for long-time fisheries professionals has been keeping up with technology, especially computers. Students graduating in 2006 grew up with computers everywhere. Students from the 1980s had some experience with computers, but very few actually owned one, and students from the 1970s had virtually no experience with computers. In the 1980s or 1990s, a district fisheries management office or hatchery may have had one personal computer that everyone shared, mainly for writing reports. By 2000, computers were smaller, cheaper, and most biologists had their own. Email replaced phone calls and letters, and floppy disks and CDs replaced filing cabinets. For the long-term fisheries professional, these were tough times because most had no experience with a computer. Over time, most professionals, some kicking and screaming, were forced to learn computers, but it became obvious to most that new technology and software made their jobs easier and made them more effective. Technological advances continue and are very evident at professional fisheries conferences. In order to be competitive, fisheries professionals must constantly keep up with the latest technologies. If you do not keep up, someone who has is going to come along and impress everyone and possibly be your next boss.

Sources of Additional Information

You should be able to find the name of the agency responsible for fisheries resources in each state in the United States at the state's official government web site (http://www.*statename*.gov). Many job openings in state agencies are posted to the AFS Job Center Online (http://www.fisheries.org/html/jobs.shtml) and others can be found with an Internet search.

References

Gabelhouse, D. W., Jr. 2005. Staffing, spending, and funding of state inland fisheries programs. Fisheries 30(2): 10–17.

Pegg, M., K. Pope, and C. Guy. 1999. Evaluation of current professional certification use. Fisheries 24(10): 24–26.

Zale, A. V., R. L. Simmonds, and R. T. Eades. 2000. Getting a job or assistantship: how to surpass the competition. Fisheries 25(6): 24–31.

Chapter 5

Fisheries Employment in the U.S. Federal Government

ROBERT L. SIMMONDS, JR. AND
MARY C. FABRIZIO

There are certainly many good reasons why a person would want to become a federal fisheries biologist (Box 5.1; throughout this chapter we use the term biologist as a generic term inclusive of scientists, technicians, etc.). The good news is that there are also many opportunities. Federal offices and laboratories that employ fisheries biologists can be found in almost every state. If you want to live in a certain part of the country or work with a particular type of aquatic system (e.g., the Great Lakes, estuaries, large rivers) or with a specific group of species (e.g., desert fishes, marine pelagics), you should be able to locate a federal office or laboratory that can provide you with those opportunities. If you cannot locate an office, share your interests with one of your professional contacts (e.g., undergraduate advisor) and ask for suggestions. Another alternative is to introduce yourself to someone at a professional meeting and ask for suggestions. The fisheries community is a small world, and if you are willing to follow up on a couple of leads you will quickly find opportunities that interest you. However, note that environmental and natural resource agencies within the federal government have different mandates, so be sure to learn about the mission of a prospective agency and determine whether your interests fit with that agency.

Once you identify particular facilities, begin a dialogue with the biologists and other staff at the offices. Let them know that you are interested in learning more about what they do, and vol-

Box 5.1. Top ten reasons to work for the U.S. federal government.

---

1. Agencies have clear missions involving conservation, protection, and enhancement of natural resources.

2. You will have days in the field that cause you to look around and say to yourself, "I cannot believe they are paying me to do this."

3. Pay is extremely competitive and cost of living adjustments occur annually.

4. Agencies provide annual training opportunities to further your skills and education.

5. A full spectrum of benefits is associated with permanent and some temporary positions.

6. You will have generous annual leave and vacation.

7. You will have 10 paid federal holidays each year.

8. Career advancement opportunities are available across the country.

9. Work schedules are flexible and there are even some opportunities to telecommute.

10. Moving between jobs, including to different agencies, is easy and your moving expenses are often reimbursed.

---

unteer if possible (everyone has *some* time to volunteer, whether it is after classes, on weekends, or during the summer). Let them know that you are ultimately interested in a permanent position, but that in the meantime you are interested in most any type of position. You can keep track of most job openings on the USAJOBS web site (presented in detail later in the chapter), but there are some positions that are never advertised. There are student positions, "special needs" or "emergency hire" positions, and others that are not posted on the web site. This is where knowing the people at an office, and making sure they know you, can pay off and get your foot in the door. Be sure to express your interest in these positions and to follow up your visit or call by sending a resume. Also, be persistent! Your first inquiry may come at a bad time during the funding cycle, and if it is your only inquiry, your desire for a position may be forgotten by the time funding becomes available.

Many of these positions have specific requirements and some are short duration positions (as little as 30 days), but they often serve as a stepping stone to better positions. They are opportunities for you to demonstrate your technical skills, work ethic, and ability to work with others. Just to be clear, these positions will not convert into a permanent position. They simply provide an opportunity for an office to get to know you. Later, when you apply for a permanent position along with 30 or 40 of your peers, you will be more than just one of many names on a list of applicants. Appropriate education and experience will get you to the top of the list, but having a connection with the office (i.e., being a tried and tested former employee) often makes the difference when the selection process gets down to the last few applicants. Working a temporary job will also provide opportunities for you to develop special skills, which will help set you apart from other applicants who are also good biologists. If an office can hire a good biologist who also has other desired skills, such as stream restoration techniques or geographic information systems (GIS) analysis, they most certainly will.

Be aware that in many federal agencies much of the work is being performed by employees working under contract or in term appointments. Contract employees need to be able to hit the ground running. Consequently, preparing for a future position by volunteering can greatly improve your chances of being hired. Term appointments are limited in duration, generally from one to four years, but they do provide the same benefits (e.g., health insurance, retirement) as a permanent position.

Volunteering, special needs appointments, and temporary or term positions all provide an opportunity to explore new areas of science and management and to determine your suitability for the tasks at hand. The types of tasks conducted at federal facilities are extremely diverse, even within a particular agency. If at first you do not find a job to your liking, continue to explore possibilities at other federal offices.

A Good Applicant

To be a good applicant, you must determine what it takes to get a job *before* it is time to find one. A solid foundation of education and experience is your first chance to set yourself apart from the competition. In today's job market, a Master's degree is a must. Several candidates will try, but very few will successfully secure a fisheries biologist position with only a Bachelor's degree. There is a large pool of applicants with Master's degrees to choose from and there is much learning and development associated with the granting of that degree. Thus, applicants with Master's degrees are far more competitive. A Ph.D. is very important for federal research positions, but it is not a necessity for the majority of fisheries biologist positions.

The quality of your education matters as well. The fact is that when a supervisor is hiring for a position there are generally a number of acceptable applicants from which to choose. Unless there is a truly outstanding candidate, the person making the decision will be looking at other characteristics, such as the courses the applicants took, their GPAs, and which schools they attended. Be sure that you can set yourself apart with regard to each of these elements. If you want to be a fisheries biologist, take as many applicable courses as possible. Be selective when it comes to choosing a college or university, particularly for your Master's degree. The quality of fisheries programs and professors varies tremendously; not all schools are created equal and some programs are more well-known or more respected than others.

Even for entry-level positions, some level of experience is expected. Of course, the standard quip applies: "Can't get a job without experience, can't get experience without a job." The easiest way to overcome this quandary is to volunteer, and the good news for federal job applicants is that volunteer and paid positions count equally. Most places you approach will be more than willing to have you work for free, so the opportunities are plentiful. Whereas most students will be busy with school and work, and will need a paying job to make ends meet, students need to keep their goals in mind. Volunteer at least a few hours or a few weeks during the summer. Volunteer experience demonstrates your commitment to your chosen field, it provides an opportunity to demonstrate your skills to potential employers, and it enhances your resume. Your toughest competition will have practical experience, so be sure you do as well. In addition to providing valuable experience and job references, your co-workers can become part of your professional network that will be *extremely* valuable throughout your career.

A good applicant will have an exceptional education, some experience in the field, or other skills that set him or her apart. Employers are looking for applicants that went above and beyond in their past jobs (e.g., received special recognition or awards) or education (e.g., enrolled in more difficult courses). Publishing your thesis or dissertation is a great way to demonstrate your willingness to follow through and to demonstrate your ability to write, a vital skill for any biologist.

Being "The One"

The successful applicant will have done the appropriate research before submitting the application and participating in the interview. Assessing the skills and talents that the employer is seeking to hire and understanding the role of the laboratory or office is key to shaping your application and guiding your interview towards the best outcome. Knowledge of the major programs or projects associated with the hiring office or laboratory is crucial. Such background information provides the successful candidate with a way to highlight his or her skills in a manner that clearly supports the vacancy.

The successful applicant is likely to possess a multitude of skills and abilities and will have experience applying those skills on the job. Federal agencies that address natural resource issues have missions that typically encompass the conservation, protection, and enhancement of natural resources and habitats. To meet this mission, fisheries science professionals in federal agencies represent many disciplines, talents, and specialties. Candidates are sought with specialized knowledge or skills, such as proficient use of statistical methods, mastery of GIS applications in aquatic and riparian habitats, command of techniques for habitat restoration, and knowledge of legislation concerning aquatic nuisance species. Demonstrated application of those skills and abilities should be evident in the application package, and may include published technical reports or articles in the scientific literature and presentations at technical conferences or workshops. Selecting officials seek out individuals with scientific or technical expertise in subject areas pertinent to the job. A successful candidate may also have demonstrated leadership capabilities and an ability to work effectively on interdisciplinary teams.

The successful applicant makes efficient use of time, has a strong work ethic, and has a track record of completing jobs correctly and on time. The degree to which the candidate has performed in these areas can be ascertained through discussions with the candidate's references. Thus, a candidate should have cultivated a supportive network of colleagues and supervisors to provide strong recommendations during job searches. This network should include professionals that have directly witnessed your performance under both good and adverse conditions (e.g., when stressed by an approaching deadline).

The successful applicant is someone that is easy to work with – a team player who is not focused entirely on advancing his or her career, but is genuinely committed to the work of the team and the mission and needs of the agency. The successful applicant gives credit where credit is due and is quick to acknowledge the efforts and accomplishments of other team members. This person is also willing to perform some of the less glamorous tasks assigned to the team because he or she knows that the work is important and must be completed. The successful applicant sets high expectations for the team, including him or herself, and facilitates the achievement of goals. The successful applicant respects and honors differences among team members, recognizing that a variety of skills and approaches are more likely to result in a better product for the team.

A candidate's communication skills are almost as important as knowledge and skills in fisheries science. Communication skills include verbal communications, but also the ability to listen. The successful candidate is able to generate good ideas and garner support for the idea from co-workers. The successful candidate is someone people want to work with, someone who is sought out for his or her ability to effectively communicate or negotiate. This person also brings fun to the workplace. On average, most fisheries professionals will work with people more than with aquatic organisms. Getting along with your supervisor, your co-workers, the public, colleagues from other agencies, or the person from whom you are buying your sampling gear is paramount to building a successful career in fisheries with the federal government.

Navigating the Federal Application Process

As the following section illustrates, the federal government has developed extensive policies and procedures for nearly every work-related task. This includes hiring and firing. For example, before a person is removed from a position for performance issues, a written plan to improve performance is developed and implemented; after a specified time, usually six to twelve months, the employee's performance is re-evaluated. Specific policies and procedures also exist for purchasing, filling out timesheets, driving a boat, traveling, and many other tasks. These policies and procedures are designed to benefit the federal government and employees in one way or another, but often create delays. The job application process can sometimes seem overly burdensome and may discourage some applicants. However, with some guidance, we show that it can be navigated successfully.

Finding a Job Vacancy

Currently, all jobs within the federal government are posted on the USAJOBS web site, which is managed by the Office of Personnel Management (OPM). On the web site you will find links to vacant positions in the federal government, including the Department of the Interior (e.g., U.S. Fish and Wildlife Service, U.S. Geological Survey), the Department of Commerce (e.g., National Marine Fisheries Service/NOAA Fisheries), the Department of Agriculture (e.g., U.S. Forest Service), and several others (e.g., U.S. Environmental Protection Agency). To effectively use this site to search for a vacancy, you should be familiar with several key features, including the job series, the type of appointment, and the pay plan or grade.

Job Series—One good way to search USAJOBS is by series or job type. Some common job types are: 0482–Fishery Biologist; 0401–Biologist; and 0408–Ecologist. If you are unsure of the job type, use key words to limit your search and then study the requirements of each series to determine which one best matches your qualifications. In addition, OPM publishes a handbook that defines the types of jobs within each series. Be aware that different agencies tend to use different job series.

Type of Appointment—Within the federal government, appointments may be permanent or temporary (including term appointments) and may be either full-time or part-time. Permanent full-time positions may include entry-level scientific positions as well as positions requiring graduate school training and extensive experience. The vacancy announcement states the type of appointment along with identification of the hiring agency and the location of the job.

Pay Plan or Grade—Many federal appointments are made according to the General Schedule (GS) grade (e.g., GS-09). Some positions are multi-grade (e.g., GS-09/11). In multi-grade positions a new employee will start at either grade, depending on his or her qualifications. For example, if the employee started at GS-09, he or she is eligible for promotion to GS-11 after successful completion of one year at the GS-09 level. Each grade corresponds to a particular salary range and is broken into ten steps. Initially, an employee moves up one step each year. Later, step increases come at two or three year intervals until step ten is reached, at which time step raises cease.

Salary tables are published by OPM each year to reflect increases due to annual cost of living adjustments. In addition, salary tables are adjusted for differences in cost of living associated with a particular location. For example, the New York–Newark salary table is adjusted upward by about 9% over the standard salary table to account for the overall higher cost of living in this metropolitan area.

In addition to the GS schedule, some agencies have started to experiment with pay banding, a method of classifying jobs that permits greater flexibility in salary ranges and promotions but does not provide the predictability of pay increases. Pay band jobs are common in the National Marine Fisheries Service/NOAA Fisheries (NMFS) and the pay plans for these positions are identified as *ZP* for scientific and engineering professional positions (e.g., fisheries biologist) or *ZT* for scientific and engineering technician positions (e.g., fisheries technician). These designations replace the GS designations. Pay banded positions have only five levels. For instance, an entry-level fisheries biologist position in NMFS may be ZP-0482-02, where 0482 refers to the series (fisheries biologist) and 02 refers to band two.

Once a job vacancy of the desired type is identified, the applicant must determine whether the position is open to the public ("all qualified applicants") or only to current federal employees. He or she must then consider the location of the job and the closing date of the vacancy announcement (in most cases this is the date by which all applications are due to the personnel office).

The applicant must examine the duties listed on the vacancy announcement to determine if his or her credentials match the requirements for the position. The duties describe the major tasks of the job and provide an overall description of the scope of assignments associated with the job. In addition, the vacancy announcement lists the specific education qualifications and experience required of a successful applicant. We encourage applicants to carefully examine the section on benefits and other information about the position. Often this is the section that describes whether the position includes locality pay and that informs applicants that only the successful applicant will be notified of the outcome (i.e., unsuccessful applicants will not be contacted once the decision is made). Consider this the fine print of the job announcement; this section can provide valuable information and may help applicants avoid disappointment.

In some cases, a vacancy announcement will include a description of special circumstances under which applicants will be considered. An applicant may be ranked higher than others because of his or her eligibility for special appointing authorities. These preferences may be afforded to veterans, persons with disabilities, Peace Corps or VISTA (Volunteers In Service To America) volunteers, displaced federal employees, persons with the armed forces, or persons with non-competitive appointment eligibility. The vacancy announcement will clearly state these restrictions. Special appointing authorities should not discourage qualified candidates from applying.

Application

Because many U.S. federal agencies use agency-specific application processes (usually Internet-based), and because these systems change frequently, we describe only the key features of the application process. Positions in federal agencies advertised through USAJOBS include specific guidance on how to apply. Sometimes the vacancy announcement includes a section that describes how the applicant will be evaluated.

Typically, the applicant will be required to respond to a series of self-evaluation questions pertaining to the job. The list of questions may be shorter than ten or longer than twenty, and the questions are usually multiple choice or yes/no. Self-evaluation questions focus on the breadth and depth of your knowledge and experience concerning essential principles or practices upon which the job is focused. For example, you may be asked to evaluate your research experience by selecting one of four responses to the following question:

Do you have work experience directing or conducting research to support the utilization, conservation, protection, or recovery of commercially and recreationally important living marine resources?

1.   I have experience managing multiple, integrated research programs and tasks.

2.   I have experience working in multiple, integrated research programs related to these tasks.

3. *I have assisted in the performance of these tasks or have worked on a temporary assignment performing such tasks.*

4. *I have no experience in this area.*

Responses to questions such as these are designed to allow Human Resources (HR) specialists to identify qualified applicants. In general, an HR specialist will determine: (1) if the applicant meets the basic eligibility requirements (e.g., U.S. citizenship); (2) if the applicant's education or background experience meets the OPM qualification requirements (these are specific to each position, and include explicit descriptions of required coursework); and (3) if the applicant's responses to the experience questions indicate that the candidate is qualified for the position. Each applicant is given a numerical rank by the HR specialist, and the ranked list of qualified candidates is forwarded to the selecting official (usually the supervisor of the position) for further consideration. Be accurate when answering self-evaluation questions. Applicants that are too reserved with their responses will not make it to the top of the list, but those who exaggerate will be found out as such by the selecting official, who will be looking for evidence to support the self-evaluation and is likely to be an expert in the field. Evidence in support of the evaluation results can be found in the resume or curriculum vitae (CV) of the applicant, college transcripts, or other written material submitted by the applicant. The announcement will specify the supporting documents that must be submitted.

Some positions require the applicant to compose responses to a series of questions pertaining to the applicant's knowledge, skills, and abilities (KSAs). Usually these questions are open-ended. For example, the applicant may be asked to describe his or her ability to initiate, develop, conduct, and publish scientific work on field and laboratory investigations dealing with fish and their habitats. We suggest writing concise, but specific responses, bringing together both educational background and on-the-job experiences to highlight the applicant's KSAs. The written narrative must be fully supported by additional documentation describing the applicant's experiences (e.g., a detailed CV). Keep in mind that you are free to submit as much additional material as you feel necessary to provide the selecting official with appropriate documentation for evaluating your qualifications. If the application does not instruct you to submit ancillary material, you may contact the HR office for instructions on how such documentation may be provided (e.g., by email, fax, or hard copy). Refrain from submitting reprints of published articles unless specifically instructed to do so.

All applicants should carefully review and edit all written material submitted as part of the application process. An application containing spelling errors, poor grammar, and incorrect punctuation speaks volumes about the applicant's ability to communicate in writing, a required skill of fisheries jobs. All written material should be consistent among documents to avoid confusion or the appearance of inaccurate or misleading information (e.g., CV indicates that your B.S. was awarded in 2005, but the narrative describing your experience indicates that your graduate studies were initiated in 2002). Inability to clearly convey your professional history may be interpreted as an inability to gather information and accurately communicate it to the public or your peers.

Interview

Interviews are considered optional and may not be part of the selection process; the selecting official determines the need to conduct interviews. For entry-level positions, the interview is typically conducted by telephone unless all of the qualified applicants are local. For mid-level and higher positions, interviews may be conducted by a search committee either in person or by conference call. Regardless of the method, prepare for the interview by reviewing the job description, job announcement, and the KSAs required of a successful applicant. Review your KSAs that address those sought for the vacancy and be prepared to give examples of how you used your skills or creatively applied your knowledge in fisheries. If the vacancy announcement indicated that the employer was searching for someone to perform fisheries surveys in large lakes using hydroacoustic technology, then be prepared to speak about your specific experiences in these areas. For example, your experience using hydroacoustic gear, your experience conducting surveys, your knowledge of fish identification and biology, your understanding of large lake ecosystems, and the skills you have for managing large data sets.

In some cases, it may be helpful to provide the names of scientists with whom you worked and to discuss funded projects in which you participated. Portray your role in the project clearly and accurately and describe how that role relates to the current vacancy.

Some jobs will require knowledge of legislative acts or policies, such as the Magnuson-Stevens Fishery Conservation and Management Act, the National Environmental Policy Act, or the Endangered Species Act. If you have no prior knowledge of these acts, it would be a good idea to become generally familiar with them before the interview. There is no need to become an expert, but at least be able to discuss the relevancy of the legislation or policies to the job.

In preparing for the interview, try to anticipate questions and formulate your responses so that you are not caught off guard or without a response. However, do not be afraid to ask for clarification if the question is vague or you do not understand. Some of the standard questions will relate to why you want the job, why you want to leave your current job, and the unique skills that you bring to the job. At an appropriate time, you should also ask questions concerning the conditions of employment, such as workplace conditions, requirements for field work or overnight travel, and whether or not the position is part of a collective bargaining unit (i.e., a union). Employees at some federal agencies are unionized, though the operation of the local unions varies greatly from place to place.

Preparing for an on-site interview is similar to preparing for a telephone interview, but you must also consider your appearance. Dress professionally (e.g., dress shirt and slacks/skirt), and remember that it is always better to overdress than to underdress. Your attire speaks to your professionalism and to your seriousness about the position. You will also likely be provided a tour of the facilities. The tour might include office space, laboratory space, and other on-site facilities such as research vessels or small work boats. If possible during your visit, speak with others in the group or on the team with which you will be working. Again, anticipate questions you will be asked and prepare a list of your own questions to help you assess whether or not you will be satisfied with the position and the conditions of employment.

At the conclusion of the interview, be sure you are provided with an anticipated timeline for the decision-making process. You should know if and how you will be contacted about the selection outcome. Some offices do not notify interviewees who are not selected, whereas others post the outcome of the selection on a web site that you can check periodically.

Selection

The selection process begins when an HR specialist provides a certificate to the selecting official. The certificate includes the names and application materials from the top three or more qualifying candidates and the ranking assigned to each applicant by the HR specialist. After appropriate review and consideration of application materials and interview results (if applicable), the selecting official completes the certificate by making his or her selection and documents the factors considered in making the selection. The completed certificate is provided to the HR specialist for review prior to being finalized. During this review, the HR specialist ensures that all applicable laws, policies, and guidelines have been observed. Once the process is deemed consistent with HR policies and U.S. federal laws, the successful applicant is notified by the selecting official or HR specialist by telephone. If the candidate accepts the position, he or she is provided with a selection letter and employment package. The selection letter is the official notification of selection and includes, at a minimum, the name of the position, the organization within which the position is found, the salary level, the start date, and the name and contact information of the supervisor.

The Successful Federal Employee

Some of you will become federal fisheries biologists, but where do you go from there? Success, whatever your definition, starts with hard work. Aside from personal satisfaction, hard work is expected and is generally noticed and appreciated by those around you. Help out on projects where your skills are needed. A "go-to" person, someone who will step up to the task and deliver, is always in demand.

Be open to opportunities. You never know when a position might come open or when your supervisor may tell a co-worker from another part of the country about the good work that you do. The world of fisheries is a small one, especially when you look within any particular federal agency. While agency offices and laboratories may span the U.S., any particular agency might have only forty to fifty facilities employing fisheries biologists. Word of a high caliber biologist can travel quickly. This could lead to opportunities when you least expect them; be sure to give full consideration to each one. Opportunities do not come in the form of job offers, as the federal government has very strict policies and procedures for assuring that hiring practices are fair to all individuals. Opportunities might come in the form of someone encouraging you to apply for a certain position. It might be your supervisor forwarding a job announcement to you from one of his or her peers in another part of the country who is looking for a good biologist. You may never have thought about living in a particular place or working for a particular office, but look critically at each opportunity to see if it is something that will help you meet your long-term goals.

Be willing to move. Fortunately, the federal government covers the cost of moving in most situations. Any individual federal office will have only a few opportunities for advancement, and very often selecting officials hire from outside the office to infuse new ideas. It is possible to advance in a given office, but you are far more likely to advance by moving. A person who is more "traveled" will have interacted with more professionals and thus should offer a larger collective experience. There will be times when moving to a new area is not possible for whatever reason, but the more willing you are to move, the more opportunities are available.

Set career goals, develop a plan, and put that plan into action. Fortunately for federal employees, one way to do this is through an Individual Development Plan (IDP) that each employee completes with their supervisor. Goals are set for successful performance in your current position as well as short- and long-term career goals. Developmental objectives are identified (i.e., what an employee needs to do in a given year to work toward the goals), as are developmental activities (i.e., training or activity needed to achieve the goals). This is an opportunity for an employee to identify career goals and to work with a supervisor who can help make those goals a reality.

Seek input on your career goals and course of action. Share your goals with others who currently hold a position you are working toward. The more you inquire, the more you will likely find that people took a variety of paths to get there, and hopefully one of those paths (or a combination all your own) will get you there. Career goals are also subject to change based on changes in your interests and opportunities that come along. The point is not to set the perfect career goals and identify the perfect path, but rather to set a target and take active steps to move toward the target.

Use the tools available to you to take an active role in advancing your career. Federal agencies encourage continuing education and some have developed their own training programs. These provide a great opportunity to take courses specifically developed for the employees of a given agency. Take advantage of the training opportunities that will help you be successful in your current position, but also those that will help you move toward your career goals. Agencies offer several leadership programs. For example, one U.S. Fish and Wildlife Service program, *Stepping Up To Leadership*, includes classroom training, completion of a team project, interviews with leaders, and shadowing of leaders. Courses like this are a great way to learn leadership skills and to determine if a leadership role is right for you. They are also a great way to get to know leaders in areas in which you want to work.

Formal training is not the only opportunity to develop skills. Identify your goal or desired position and identify the qualities or experience necessary for success in that job. Use that information to begin acquiring those qualities or experiences by volunteering to assist someone in that position with a project, reading about the projects they direct, or pursuing a temporary assignment to their office. Take steps every month to prepare yourself for that position.

Finally, remember to build and use your professional network to achieve your career goals. Talk to your supervisor or other people in your network about your goals and how to achieve them. Ask them to put you in touch with people in the positions you seek who can help you better understand and prepare for the role. Get to know the leadership in your agency; it may be intimidating at first, but good leaders are glad to talk to you and are very interested in identifying and helping the leaders of the future. Build a broad network of contacts and relationships that will help you prepare for the role and will help you secure that role when it becomes available. You never know which opportunity or which contact will be key in helping you reach your goal, so be sure to seize opportunities and build your network whenever you can.

Sources of Additional Information

Federal job announcements are posted on the USAJOBS web site (http://www.usajobs.opm.gov), which is maintained by the Office of Personnel Management (http://www.opm.gov). The Office of Personnel Management publishes the *Handbook of Occupational Groups and Families*, which defines the types of jobs within each job series, as well as the federal salary tables. Both documents can be found by searching the OPM publications on their web site.

Chapter 6

Academic Positions in Fisheries Science

BRIAN R. MURPHY

Your research is completed, your dissertation is in close-to-final draft, and your defense is scheduled. Now you start to wonder, "Where do I go from here?" Many people find the idea of an academic position at a college or university to be attractive. This chapter focuses on academic faculty positions, but also briefly discusses postdoctoral research positions in a separate section at the end of the chapter. University or college professors typically divide their time between teaching, advising students, participating in the work of committees and administration, seeking research funds, conducting research, performing public outreach, and writing scientific and academic publications. Time spent on each of these tasks varies widely, depending on the type of academic position and institution.

Before you apply for academic positions, be sure that you understand what is expected of fisheries scientists in these positions. Most academic positions are tenure-track slots, meaning that the candidate serves a probationary period (usually 5 years) as an Assistant Professor, at which point they are evaluated for promotion (to Associate Professor) and tenure (life-time appointment). Failure to achieve promotion and tenure generally means termination from employment with that institution, typically after a 1-year grace period given to find a new position. Requirements for promotion and tenure vary between institutions, but generally include the development of an effective teaching and research program. Success is measured by student and peer evaluations of teaching; success in securing research grants, completing research projects, and publishing research articles in peer-reviewed scientific journals; and successfully directing graduate students to completion of their degrees. If you continue to perform at a high level for a number of years, then further promotion to the rank of full Professor is possible.

Some people feel that the pressure to perform makes this a high-stress career (e.g., the "publish-or-perish syndrome," particularly in the early years), but many professors enjoy the intellectual stimulation and the great variety of daily challenges in the job. The description above applies primarily to major research universities (also called Research I Institutions). If you prefer more emphasis on teaching, then a smaller "teaching school" (e.g., liberal arts college) may be more to your liking. Advancement at these institutions traditionally is based more on the quality of teaching than on research success, but these institutions also are beginning to have significant expectations for research. These two extremes are the bounds of a wide universe of academic program types, and you need to make sure you understand the role of faculty and the expectations for success at any program to which you might apply. For instance, some Research I institutions have fisheries-related programs that are graduate only, so teaching expectations would be light and only at the graduate level. Other Research I schools have both undergraduate and graduate programs for fisheries students and employment expectations may involve some teaching at both levels. A candidate's success at these Research I institutions will be based primarily on research productivity. On more middle ground, there are many comprehensive universities, which are broadly defined as those institutions with diverse undergraduate offerings, multiple M.S. and M.A. programs, and few, if any, Ph.D. programs. Teaching is often very important at comprehensive schools, but teaching loads are usually not so heavy as to prevent one from establishing a moderately sized research program. Tenure and promotion decisions at these institutions can be based more equally on teaching and research. Finally, at the opposite extreme from Research I institutions, there are many undergraduate teaching universities and colleges that focus solely on teaching, where heavy teaching loads can be expected. At each institution, however, there is bound to be considerable variation among departments, and the candidate should do their best to obtain a clear understanding of the expectations for success specific to the department.

Not all faculty positions have similar job duties. Depending on the mission of the college or university and the specific type of position, faculty may be involved in teaching, research, extension (public outreach), or any combination of the three. Extension positions in particular may be outside the realm of experience for some job applicants. Extension positions are generally located at land-grant institutions, and extension faculty have responsibility for developing various types of public education programs to take the results of research to appropriate user groups. Extension faculty may be subject to the typical academic review for promotion and tenure, or their system for advancement may be outside the typical promotion and tenure process.

In addition to typical tenure-track professorial positions on the major college campuses, there are many other types of academic faculty appointments, particularly in the fisheries field. Some colleges and universities maintain coastal research labs separate from the main campus, where faculty could be employed in any of the areas of teaching, research, or extension. Positions at these labs offer many obvious advantages by nature of being closer to the resources under study, but sometimes these labs and positions are outside the mainstream of academic life (and opportunities) on the main campus.

Some university faculties also include various governmental scientists who are assigned to a college campus. Perhaps the best known of these government programs is the Cooperative Research Units Program of the U.S. Geological Survey (USGS; see Chapter 7), but there are other similar appointments in some state and federal agencies, and even with Native American tribal groups. Examples of these additional programs include university-based scientists from the U.S. Forest Service, the National Park Service, the National Marine Fisheries Service, the Michigan Department of Natural Resources, the Illinois Natural History Survey, and the Nez Perce Tribal program at Washington State University. Objectives of all of these programs vary, but usually include some level of specialized research and teaching. Since most scientists in these positions hold joint appointments as university faculty, the host university will have a major role in the hiring or appointment of scientists in these positions. In that sense, much of what is discussed in this chapter will also hold for these positions, but there will be many differences and exceptions. If you seek one of these specialized governmental appointments, be sure to investigate the specific role that these faculty play on campus and determine any specialized hiring requirements for these positions.

Talk with professors and other academic professionals about their jobs to help you decide if this is the career you wish to pursue and what type of school would suit you best. If the life of a professor appeals to you, then it's time to start a search for a suitable academic position.

Locating Positions

Unlike some types of positions in private industry and government, the shotgun approach to finding employment—flooding the world with your resume—rarely works in academia. In these days of tight budgets and tightly controlled hiring practices, colleges and universities can only hire new faculty to fill specifically advertised vacancies. Watch for vacancies at the AFS Job Center Online (and related listings in *Fisheries* magazine), in *Science* magazine and *The Chronicle of Higher Education*, or posted on bulletin boards when your department receives direct mail announcements. Many openings are "announced" by word-of-mouth before advertisements even appear. Throughout your career you should have been building a network of colleagues in fisheries, both at your own and at other institutions. Make sure that your colleagues know that you are seeking an academic position so that they can refer appropriate announcements to you.

Collecting Background Information

Once you have identified an open position in which you may be interested, you should begin to collect as much information as possible about the college or university, the program, and the faculty in the department to which you will apply. Start on the Internet, or at your library's reference department. Browse through the undergraduate and graduate catalogs from the candidate school. These can be a good source of general information about the school and the features of the surrounding area. Investigate the scholarly resources of the campus library. Familiarize yourself with student groups affiliated with your target program. Look at the degrees and options offered by the department, and the courses that faculty teach. Next, consult the program's web site to obtain information on the program's faculty, their educational backgrounds, and their research interests. If you want further information on research of individual faculty, check to see if they have a web page and search the literature for their publications through various indexes and abstracting services (e.g., Web of Science, Aquatic Sciences and Fisheries Abstracts, Google Scholar). Be particularly familiar with the work of faculty members who are most likely to work closely with the successful candidate who secures the job. For information on faculty salaries, check the Faculty Salary Survey conducted by the American Association of University Professors (AAUP), as well as the AAUP's Annual Report on the Economic Status of the Profession. *The Chronicle of Higher Education* also has a wealth of information about faculty salaries and related topics, including their annual Almanac. The National Education Association (NEA) also publishes an annual Almanac, which includes a survey of faculty salaries. For information about the cost of living in the area, K-12 school system ratings, crime rates, climate, recreational opportunities, and other general information for many metropolitan areas of the United States, check the current *Places Rated Almanac*. Local newspapers with online editions can also be an excellent source of information regarding local issues and the cost of living in the area.

Feel free to contact personal acquaintances that you know in the department to which you will be applying. This can be a good way to get the "inside scoop" on exactly what the faculty hope to find in a new colleague, but do not expect your acquaintances to provide information that they could not provide to every caller. If you do not know anybody in the department, contact the search committee chairperson for answers to questions you might have about the position or the program, but do not become a pest throughout the search process.

The Application Process

What to Send

Read the job announcement carefully and follow all directions for submitting an application. Send only those materials that are requested, which generally will be a statement of interest in the position (i.e., cover letter), a complete curriculum vitae (CV), and a list of persons who have agreed to serve as professional references (including their phone numbers, email addresses, and an indication of your relationship to that person). Do not have references send letters of recommendation unless the job announcement specifically states that you should do so. Often the search committee will make a first cut of applicants before requesting letters of recommendation. Typically, the search committee is faced with going through dozens or even hundreds of applications; try to make their job as easy as possible. Avoid sending a large packet of reprints that nobody is likely to read closely and never send more than a few of your best publications if you are instructed to send reprints. A brief statement of teaching and research interests often is requested or can be added at your discretion; use this statement to demonstrate a logical, well-reasoned evolution of your philosophy and approach to teaching and research. All materials that you send should be complete yet concise, well-organized, and well-written. Make certain that you check for typographical errors and misspellings several times before you mail your application. Ask your major professor or other colleagues to review your materials for accuracy and clarity before you submit them to the search committee.

Your CV

Your CV should be in an easy-to-read format that provides all of the information that will be critical to the search committee as they make their decision (see Chapter 2 for a detailed discussion of resumes and CVs). There are some important points to consider relative to your CV for an academic application. As you customize your CV for the position to which you are applying, think about the varied tasks that faculty members perform and show how your experiences and accomplishments demonstrate your potential for success in the position. Do not pad your publication list with all types of reports, presentations, abstracts, and other non-refereed items. Show refereed publications in a clearly labeled section and put other types of publications either in their own sections or in a general "Other Publications" section. List all co-authors on all publications, in the proper order of authorship. Limit listings of personal information to only those things that will have some bearing on your qualifications for the job. Avoid flamboyant paper colors, fonts, and graphics for your CV. Do not let your experiences and accomplishments be overshadowed by aggressive marketing ploys that may turn off some people.

The Interview

The search committee will sort through all of the applications that are submitted and generally select two to four candidates for interviews. You may be asked to interview with very little advanced notice, so be preparing for an interview before you actually are contacted (see also Chapters 4 and 5). You can be sure that you will be asked to present a seminar during any interview, so begin preparations early. On-campus interviews may be preceded by a telephone interview with the committee; such a phone interview also requires careful preparation on your part, as it can be important in securing an invitation for an onsite interview (Campbell et al. 2002).

Meeting People

The impression that you make on people when you first meet them might turn out to be one of the deciding factors in the selection process, so do your best to create a good impression. Dress appropriately (coat and tie for men, equivalent level of formality for women), and try to be friendly, but not pushy. Be aware that your body language may sabotage an otherwise successful interview. Shake hands firmly with people you meet; maintain confident, but not glaring, eye contact; and be sure to use natural, erect posture both in sitting and standing. Expect to be scheduled to meet with the search committee, various faculty members, and any number of administrators from the dean to possibly the president. Be prepared to carry the conversation in some of these meetings, as busy faculty and administrators may not have taken time to re-read your CV before meeting with you. Expect to be asked to talk about your background and interests. Be prepared for questions like "Why do you want to come here?", "What do you perceive as your greatest strengths and weaknesses?", or "What is the first thing you would do if you were hired?" Be honest, but keep your answers positive, as you do not want to come across as hard to please or a complainer. Be prepared to discuss the types of classes you would be interested in teaching and research interests that you would plan to pursue. Be ready with questions of your own about the institution and the local area to keep the conversation from lagging.

To be well prepared, you might develop sets of specific questions for the dean, the department head or chair, other faculty members, and students. There are many possible questions to include, but general topics might include the college's or university's mission, administrative support, faculty mentoring, student quality, available teaching and research support, outreach expectations, available space, tenure expectations, and salary. You want to project an image that you understand the nuances of academic positions and the sociology of science in the modern academic institution. Work with your faculty mentors to prepare these lists of questions for various people or groups that you will encounter during the interview. Having a list keeps the conversation from going dormant and a typed list of questions projects an image of preparedness. You also might be scheduled to meet graduate and undergraduate students, departmental staff, and even a local real estate agent; if not, ask whether these meetings might be possible, as each of these can give you valuable insight on the institution and the local community.

The Research Seminar

Undoubtedly you will be asked to present a seminar about your research for faculty and students. This is not just an easy way for faculty to learn about your research interests, but is also a test of your communications skills and potential as a classroom teacher. The importance of thorough preparation for this presentation cannot be overemphasized. Only by doing extremely well with this presentation can you have any hope of landing the job.

The research seminar is the occasion within your interview where you will have contact with the largest number of people and for some of them this may be their only exposure to you. Various people will use the seminar, rightly or wrongly, to judge you on everything from your personality and collegiality, to your organizational and communications skills, to the quality of your research and your prowess as a teacher. For this reason, you need to prepare the very best seminar that you possibly can. A few critical points to consider include the following:

1. Gauge the audience that can be expected to attend your seminar, and be sure to have something for viewers at every level.

2. Do not try to amaze everyone with your technological or statistical prowess. Keep your slides simple and clear and limit statistical discussion of your data to only the critical points for understanding.

3. Present a balanced seminar that shows both depth and breadth in your abilities.

4. Be sure that your slides are free of errors and definitely stay within your allotted time. Rehearse your presentation numerous times to mixed audiences of colleagues. Be particularly comfortable with your introduction and conclusions. Ask colleagues for truly critical evaluations and take their sugges-tions seriously.

5. Anticipate questions that are likely to be asked. Be prepared to answer the difficult ones with confidence and straight talk. Your answers should be concise and end on a positive note.

The Teaching Seminar

Some interview schedules include an additional type of seminar. Do not be surprised if you also are asked to present a guest lecture in an undergraduate class, or a second seminar to simulate such a teaching opportunity. This is a fairly common practice at teaching colleges and comprehensive universities, and is becoming more common at larger research universities. This seminar provides the search committee with an appraisal of your teaching skills. Be sure that your teaching seminar is just that; do not try to simply recycle a research seminar that you have used elsewhere. Make your teaching seminar a true classroom presentation that you would present to an audience of students. Be sure to ask what level of students your teaching seminar is meant to address. Start with some specific learning objectives, present a concise and logically sequenced case, and be sure to summarize the major points at the end.

The Exit Interview

Often your last meeting will be with the department head or chairperson. This will be your chance to discuss your expectations and those of the institution, if these have not come up already. Be prepared to talk about potential research collaborations that you may have discovered during the interview. Be sure you have a clear understanding of what space, equipment, and start-up funding will likely be available for the successful candidate. Also, clarify the institution's and department's expectations for teaching load and accomplishments needed for tenure. If you are not asked about salary expectations, bring up the subject. You should expect a salary level near the median of faculty at the same rank in which you would start. Be sure you understand whether the position is a 12-month appointment or something less. The interview will likely close with a discussion of your views on the program and the potential position. Keep all of your comments positive, as negative comments about facilities, location, or departmental programs, faculty, or students will set a sour tone for your visit. Emphasize the positive if you want to stay in contention for the position. No department wants to hire someone who is complaining about things before they even have the job!

Postdoctoral Research Positions

Postdoctoral research positions are an increasingly common academic opportunity for new Ph.D. recipients. Typically these are short-term (1 to 3 years) research positions in laboratories of established college or university professors. As research budgets have grown and demands on faculty members' time have increased, some professors have turned to hiring postdoctoral research scientists to help ease the load. These are not career positions, but rather a career development step between that of graduate student researcher and full-time faculty member. These positions offer a chance to further hone your research and publishing skills in an environment that includes more responsibility for project design, funding, oversight, and publication than a typical graduate program. Postdoc positions also can offer interim employment while you seek a permanent position, and such postdoctoral experience may even be listed as a requirement for some advertised faculty positions. It will be up to you to decide whether you want to spend an additional few years in such a temporary position before you move on to permanent academic employment.

Conclusion

Preparation and planning are the keys to a successful academic job search. Plan your graduate program to provide the experiences that will make you an attractive applicant. Do your homework before applying and before interviewing. Going to an interview well prepared will allow you to relax, be yourself, and make a good impression.

Sources of Additional Information

Data from the AAUP Faculty Salary Survey can be accessed through the web site of *The Chronicle of Higher Education* (http://chronicle.com/stats/aaup). The AAUP's Annual Report on the Economic Status of the Profession can be found at their web site (http://www.aaup.org). The Facts and Figures section of the web site of *The Chronicle* (http://

chronicle.com/stats) provides access to a host of other sources of information about faculty salaries and related topics, including *The Chronicle*'s annual Almanac of Higher Education. The annual Almanac of Higher Education published by the NEA can be accessed through their web site (http://www.nea.org).

Suggestions for Additional Reading

Campbell, R. W., M. C. Horner-Devine, J. Lartigue, and G. C. Rollwagen Bollens. 2002. Preparing for an academic job interview: frequently asked questions for on-site and phone interviews. American Society of Limnology and Oceanography. Available: http://aslo.org/phd/interviewhints.pdf. (March 2006).

Heiberger, M. M., and J. M. Vick. 2001. The academic jobs search handbook, 3rd edition. University of Pennsylvania Press, Philadelphia.

Stasny, E. A. 2001. How to get a job in academics. The American Statistician 55: 35–40.

Chapter 7

Fisheries Employment in Cooperative Research Units: Where Agency Meets Academia[1]

THOMAS J. KWAK AND
F. JOSEPH MARGRAF

Many graduates in fisheries science and related fields flourished in the learning environment of their university and would like to continue enjoying that atmosphere, while others seek the close interaction with aquatic resources and fishery constituents that is experienced in natural resource agency employment. If you are hesitant to give up the flexibility and stimulation of campus life, but would like to make a direct and positive impact on fisheries and aquatic resource management, a position in a Cooperative Research Unit may be your place in the profession. Cooperative Research Unit employees work on university campuses and are part of the academic community, but conduct research to address the needs of natural resource management agencies. Interactions between Unit employees and agency biologists are an important function of each unit. The primary employment opportunities at Cooperative Research Units are Unit scientist positions (Unit Leader or Assistant Unit Leader), which are permanent federal positions with the U.S. Geological Survey (USGS) in the Department of the Interior. These positions require a doctoral degree. Other Unit positions may be available at Doctoral, Master's or Bachelor's levels of education; these are usually administered by the host university, are grant-funded, and are of varying duration.

---

[1] The views expressed in this chapter are those of the authors and are not presented as the official position of the U.S. Geological Survey or other cooperating agencies or institutions.

What are Cooperative Research Units?

Cooperative Research Units are each a joint venture among the federal government (USGS), a state natural resource agency, and a state host university (usually the land-grant university). The Wildlife Management Institute, U.S. Fish and Wildlife Service, and other state and federal agencies may also serve as cooperators at various units. The three-point mission of Cooperative Research Units is *Research* on renewable natural resources to address the information needs of cooperating agencies and partners, *Education*, primarily in the form of graduate student mentorship and teaching, and *Technical Assistance* and training to cooperating agencies and others toward improved natural resource management. Cooperative Research Units provide the mutually beneficial connection between natural resource agencies and universities. Federal and state agencies are afforded access to Unit scientists, other university faculty, and campus facilities, while the host university benefits by additional expertise on campus and access to relevant agency research and training opportunities.

The first unit was established in 1935 at Iowa State College by visionary J. Norwood "Ding" Darling. His pioneering efforts were in response to the human changes in land-use practices and associated declines in renewable natural resources of the time. Another goal was to provide trained individuals in government to guide management of these resources. Darling's concept of a nationwide network of Cooperative Wildlife Research Units was realized that same decade as a federal program of nine Wildlife Units. The program gradually expanded among states over time, and in 1960 Congress gave statutory recognition to the program by enactment of the Cooperative Research Units Act, providing a line item for the program in the federal annual budget. While the congressional act improved the program's stature and stability, perhaps the most important provision of the act was to expand the program to include fisheries at Cooperative Units, and the development of several Cooperative Fishery Units followed. The Cooperative Research Units Program has a rich and interesting history that is eloquently described and detailed by W. Reid Goforth in a document that celebrated the 70th anniversary of the program (Goforth 2006).

Today, there are 40 Cooperative Research Units in 38 states (Figure 7.1). Most are combined Fish and Wildlife Research Units (33 units), but several separate Fishery Units (5) and Wildlife Units (2) also exist. The program is staffed by more than 110 Ph.D. scientists who may advise as many as 600 graduate student researchers. Each Unit is staffed according to the unique needs of its host state and university. Additional states have expressed interest in establishing new Cooperative Research Units, and the program is expected to grow.

Research topics of Unit scientists can range from those of local or state interests, to those of regional, national, or international scope. Research is directed toward the needs of the resource management agencies, usually as appropriate for involvement by graduate students for incorporation into theses. Funding is often provided by the state or federal agency requesting the work but sources can include nongovernmental agencies, corporations, and private foundations. Unit students may also be funded by traditional academic sources (e.g., fellowships, teaching assistantships), but most are supported by research assistantships associated with research and grant funding initiated by a Unit scientist. Research conducted by Unit scientists must be approved by a coordinating committee composed of the Unit cooperators.

Unit Employment

The typical Fish and Wildlife Unit is staffed by three federal scientists: a Unit Leader and two Assistant Unit Leaders. Two scientists (one Leader and one Assistant Leader) are staffed at Fishery Units and Wildlife Units, and a few states support Units with four or five scientists. The Unit Leader may be trained either in a fisheries or wildlife discipline, whichever is deemed appropriate by the cooperators to meet the needs of the Unit. The Unit Leader is a senior scientist who has usually served as an Assistant Unit Leader somewhere in the Unit program. In addition to his or her scientific duties toward accomplishment of the Unit's mission, the Leader is administratively responsible for the activities of the Unit. If there are two Assistant Unit Leaders, one is usually trained in fisheries and one in wildlife, and their primary duties are directed to accomplishment of the Unit's mission. Assistant

Figure 7.1. Locations of 40 Cooperative Research Units at host universities in 38 states.

Unit Leaders may be at any stage in their professional careers from entry-level to senior scientists. New or recent Ph.D. graduates can expect to compete for positions as Assistant Unit Leaders. Most Unit scientists are trained in fisheries or wildlife disciplines, but scientists with expertise in other areas may be hired as well to meet cooperator needs. Scientists trained in areas related to natural resources, such as ecology, behavior, toxicology, physiology, genetics, statistics and modeling, spatial analysis, and human dimensions have been employed at Units.

Unit scientists (Leaders and Assistant Leaders) must hold a Ph.D. degree because they serve as faculty at host universities, but opportunities also exist at units for other researchers with Ph.D., M.S., or B.S. degrees. Unit scientists are known for success in securing extramural funding from cooperators and other sources, and this funded research activity requires recruiting and hiring researchers at all levels of education in addition to graduate students. Of course, the longevity and security of such temporary staff positions depend on the funding amount and duration of the grant. In some cases, however, temporary research staff have held positions at Units for many years, supported by a series of funded grants. Further, opportunities are available for postdoctoral or post-Master's researchers to develop proposals and seek funding to support their position in collaboration with their Unit scientist supervisor. Many Units also employ full-time or hourly technical assistants (usually undergraduate students or recent B.S. graduates) to organize and repair research equipment, maintain vehicles, and assist in the field or lab, and Units often host students or recent graduates in internships. All of these temporary Unit positions can serve as professional "stepping stones" to improve one's qualifications for subsequent opportunities for graduate school or permanent employment.

By and large, the most prevalent temporary positions at Coop Units are graduate research assistants who are working toward an advanced degree. Overall, the graduate student experience for those students advised by Unit scientists is not much different than that for other university graduate students in the same department or school (see Chapter 3). Three primary but subtle differences exist, however. First, as supervisees of a Unit scientist, Unit graduate students technically serve as volunteers of the federal government and as such are required to complete specific safety training for some field procedures (e.g., boat operation, electrofishing, low-altitude flying). In addition to facilitating student education and improving safety, such training and certification enhances the student's qualifications for future employment. Secondly, Units tend to accumulate research equipment and vehicles through cooperator support that are an asset to students initiating thesis research. The availability of federal vehicles and boats is typical for Unit students, but the inventory of other research equipment relies on grant funding awarded to Unit scientists that serve as student advisors. Thirdly, the Unit graduate student experience nearly always involves close interaction with agency biologists, which can be enlightening and useful for future professional development. This interaction is usually accomplished by virtue of research funding provided by an agency, but also occurs through the expectation of both Unit scientists and cooperating agency biologists to partner in achieving all three Unit mission components (research, education, and technical assistance). Agency biologists may return to graduate school for additional training at a Unit with the support of their agency, which can improve the atmosphere, diversity, and educational experience at that Unit and university. Of course, many other university departments and faculty advisors may function in similar ways as Units and their scientists to provide training, equipment, and agency interaction, but these are common features that you should expect to encounter at Coop Units.

The Ups and Downs of Unit Positions

Some say that Unit scientists are federal employees stationed on a university campus, whereas others refer to them as university faculty members that are salaried by the federal government, but the fact of the matter is that both perceptions are true and neither alone is a complete description of a Unit scientist position. Unit scientists are employees of USGS housed on university campuses *and* they serve as faculty members and graduate student advisors in a home department or school. Thus, Unit scientists enjoy the best and endure the worst aspects of both agency and academic employment.

Unit scientists are both federal employees and faculty, so they have all the resources and facilities of the federal government and their host university available to pursue their work. On the other hand, they must produce results that support three institutional missions (federal, state, and university), follow multiple sets of rules and guidelines, and answer to three bosses. An important advantage of Unit positions is that the scientist can wear either the agency hat, the academic hat, or both, depending on the situation. They also enjoy most all faculty privileges and university resources (e.g., administrative and grant support, laboratory facilities, university motor pool, library services, athletic facilities, access to teaching assistantships and other university support), as well as access to the vast assets and support of the federal government (e.g., streamlined processing of federal grants, access to government travel resources, surplus equipment, and vehicles). Unit scientists may have access to funding opportunities reserved for agency scientists, but may not be eligible for certain university funding sources, such as new faculty grants. One privilege that separates Unit scientists from other federal scientists who may be located on campus is that Unit scientists may serve as principal investigators on university grants.

Wearing two hats, however, may require double the amount of paperwork and other minutiae that most scientists seek to avoid. Unit scientists are required to complete numerous federal training modules in areas such as ethics and diversity, computer security, and workplace whistle-blowing, in addition to courses in field safety procedures. Published works of Unit scientists must undergo agency review for potential policy implications, in addition to the usual peer review for scientific content. Unit scientists must also follow university regulations and training for animal use and care, laboratory safety plans, and other university administrative or safety requirements. Assistant Unit Leaders undergo annual federal evaluation by the Unit Leader and Unit Leaders are evaluated by their regional Unit Supervisor. All Unit scientists participate in four-year panel reviews of their research productivity and all participate in departmental evaluation by their university. Other grant-funded Unit researchers (postdoctoral, post-Master's, and other staff) are university employees and avoid the federal administrative burden, but must meet all federal training requirements, and are evaluated by their university supervisor, usually a Unit scientist.

How do Unit scientist positions compare to tenure-track university faculty positions? No two Cooperative Research Units are alike, and the same may be said of universities and their departments, but several fundamental differences exist between typical Unit scientist positions and tenure-track faculty positions (Table 7.1). In general, relative to faculty positions, Unit scientist positions emphasize research over teaching, extension, or outreach. Salaries and benefits are usually comparable, but Unit scientists may not financially participate in professional consulting. Relationships with agencies facilitate grant funding, job security is comparable, evaluation procedures are functionally similar, academic freedom is more limited, agency interaction is greater, and prestige and respect are generally similar. The Unit Leader position includes a substantial administrative component, and if the individual is not released of other duties or recognized for that additional responsibility, then this can serve as a disadvantage over the Assistant Unit Leader position. Unit Leaders usually enjoy added prestige, shape cooperator relations, and influence the atmosphere and focal directions of the Unit, which may be rewarding, but there is no salary supplement associated with the position.

Table 7.1. Comparison of job attributes typically associated with Cooperative Research Unit scientist and university tenure-track faculty positions.

Job Attribute	Cooperative Research Unit Scientist	Tenure-Track Faculty Position
Primary responsibilities	Research emphasis, teaching and extension reduced; graduate education emphasis; administrative component for Unit Leaders	Variable among positions, substantial research and teaching or extension emphases at undergraduate and graduate levels typical
Salary	Federal GS scale, includes supplement for some areas with inflated costs of living; 12-month employment; annual raise most years, increases with GS-scale promotion; consulting for additional pay prohibited	Variable among universities and positions; appointment may provide 9 or 12 months salary, with potential for grant supplement; annual raises fluctuate with state budgets, increases with tenure and academic rank; consulting for additional pay usually allowed or encouraged
Funding	Advantage with cooperating agencies, availability of state and federal operating funds, start-up funding minimal, restricted from certain competitive programs	Advantage with university fellowships or funding programs, start-up funding usually substantial, no restrictions on competition
Job security	Generally secure after 1 year probation, but subject to cooperator approval and federal budget	Low security until tenure review, then high if granted
Evaluation and promotion	Reviewed by supervisor annually, reviewed by one or more federal panels for promotion, focused on research accomplishments	Reviewed by faculty colleagues and university administrators, according to research, teaching, or extension components of appointment

The history, notoriety, and camaraderie of the Cooperative Research Units Program are of great benefit to Unit employees. Being part of a larger program serves Unit scientists in many ways. Morale is boosted by the interaction with scientists at other Units; this

Table 7.1. Continued.

Job Attribute	Cooperative Research Unit Scientist	Tenure-Track Faculty Position
Prestige and respect	University appointment (adjunct or regular faculty) and respect variable among Units and institutions	Generally high, but may vary with department or appointment
Agency interaction	Substantial and expected	Variable at individual discretion
Academic freedom	Activities must be approved by cooperators, policy implications reviewed by administrators; federal regulations and university conduct standards must be followed	Freedom to pursue most topics within university standards of conduct

occurs spontaneously or through planned workshops or occasional meetings of federal personnel from all Units. The administrators and staff of the Cooperative Research Unit Headquarters in Reston, Virginia, provide constant and strong support for Unit scientists, staff, and students, and opportunities for scientific collaboration are great. Examples of such collaboration include nationwide initiatives, such as the GAP Analysis Program to develop Geographic Information System layers for habitat and biodiversity for each state (Scott et al. 1993), and a regional collaboration among Unit scientists from states bordering the Missouri River to study the population structure and habitat use of benthic fishes of that system on a broad spatial scale (see http://infolink.cr.usgs.gov/Science/BenthicFish).

How to Land a Unit Job

Openings for Assistant Unit Leader positions are usually advertised in much the same way as university faculty positions. The application process, however, is different from that for a faculty position because Unit staff are federal civil service employees. The applicant must submit an application to the appropriate regional government personnel office (as indicated in the job announcement) and should be sure to completely address the ranking factors of Knowledge, Skills, Abilities, and Other Characteristics listed on the job announcement (KSAs; see also Chapters 4 and 5). The personnel office certifies the eligibility of the applicants and forwards two lists of eligible candidates for further consideration by the cooperators. One list includes only those applicants who are current federal employees, and the other lists all other eligible candidates. Certain military veterans receive special consideration in the process. From these lists, the cooperators will select their Assistant Unit Leader, using an interview process similar to that for a university faculty member (see Chapter 6). The individual is chosen by consensus of all the Unit cooperators, but the host university and state agency cooperators are typically afforded more influence.

There are several suggestions that applicants can follow to enhance their chances for success in attaining a Unit position. First, follow directions and closing dates exactly for the federal submittal, as any deviation or late applications will disqualify your application. Be extremely thorough in describing your skills and credentials, including even basic skills and accomplishments that biologists may take for granted. If you are in doubt about the application procedures, you should call the personnel office in the announcement. It is also wise to contact the corresponding Unit Leader to express interest and find out more about the opening. It may be beneficial to provide to the Unit Leader an additional packet of information that may include a cover letter summarizing your interest and special qualifications, a curriculum vitae, and any other helpful documents (e.g., representative publications). This will make the Unit Leader and local cooperators aware that you applied to the federal office, because staff in the personnel office are not scientists, and they may not prioritize the list of candidates the same as those scientists in the Unit or cooperating parties. If you are invited to interview, you should be prepared to visit with cooperating agency administrators or biologists, as they will participate in the selection process. Also, be sure to prepare your seminar with both academic and agency audiences in mind. Familiarity or experience with the Cooperative Research Units Program will be viewed favorably by cooperators.

If you are selected for and offered an Assistant Leader position, the degree of negotiation for specific employment terms is limited. The federal grade level (salary) is fixed unless you have substantial postdoctoral experience, and then is only slightly flexible. You should generally expect only modest start-up funds, but the federal government provides some moving expenses even for new employees. At most Units, the state cooperating agency is eager to work with new Unit scientists, and a grant award from them for a specific research project is a reasonable expectation. Other grant-funded hiring at Coop Units follows typical university procedures and protocols and involves direct contact with the hiring Unit scientist (e.g., graduate students; see Chapter 3).

The Cooperative Research Units Program is committed to improving diversity among scientists in the fisheries field. Efforts toward this end include consideration of demographics in filling Unit scientist positions and in admitting graduate students to pursue studies under the direction of Unit scientists. One initiative to facilitate the education and employment of underrepresented groups by the federal government and USGS is the Student Career Experience Program (SCEP; see also Chapter 14). Students in SCEP may receive tuition or stipend assistance while in school and then may be non-competitively hired in a federal agency within 120 days of completion of their academic program. The intent of the program is to improve diversity in federal agencies, but anyone is eligible to participate and Unit scientists and students have used this program.

Parting Advice

Cooperative Research Units blend agency and academic goals, perspectives, and work environments, both in positive and negative aspects. Unit scientist positions are well suited for those who are comfortable spanning both environments, and they can be extremely rewarding. It is worth noting that other federal and state agencies have developed similar cooperative units and fisheries positions on select university campuses (e.g., U.S.D.A. Forest Service, National Marine Fisheries Service/NOAA Fisheries), but the USGS program is the most permanent and comprehensive. These positions are challenging, intellectually and personally. They typically require substantial interpersonal communication, and disappearing to your office or laboratory is not a possibility. They are also not a means to avoid tenure review in the academic setting; Unit scientists are reviewed for promotion to the same standards as faculty in their home department.

Hiring in Cooperative Research Units tends to be cyclic and tracks federal budgets, so it is important to keep informed and look for vacancies constantly, as several may be advertised at once following a prolonged period of no hiring. When considering which Units are of interest, young professionals should bear in mind that transferring to a position in another Unit is possible after attaining an Assistant Leader position. Such transfers are not simply non-competitive relocations, and Unit scientists must apply and compete for vacancies at other Units, but a Unit scientist is usually at a competitive advantage based on their knowledge and experience. Conversely, Unit scientists may be less competitive for university tenure-track faculty vacancies in some cases, as they may be perceived as agency scientists.

No two Units are alike; they vary with resources, personalities, and dynamics among cooperators, so it would be wise for prospective scientists to do their homework to understand the nuances of individual Units of interest. The best approach to learning more about employment in the Cooperative Research Units Program is to interact with scientists currently in those positions and gain insight from their experiences. Most will be willing to share a candid discussion with you on their position, those at other Units, and how to achieve your professional goals.

Sources of Additional Information

General information about the Cooperative Research Units Program and links to individual Unit web sites are available on the program's web site (http://www.coopunits.org). You can also obtain additional information by contacting:

Chief, Cooperative Research Units
U.S. Geological Survey
12201 Sunrise Valley Drive
Mail Stop 303
Reston, Virginia 20192
Phone: (703) 648-4260
Fax: (703) 648-4269

Information about opportunities with USGS through the Student Career Experience Program is also available online (http://www.usgs.gov/ohr/student). Federal job announcements, including Unit scientist vacancies, are posted on the USAJOBS web site (http://www.usajobs.opm.gov; see also Chapter 5).

References

Goforth, W. R. 2006. The Cooperative Fish and Wildlife Research Units Program. Special Publication, U.S. Geological Survey, Cooperative Research Units, Reston, Virginia.

Scott, J. M., F. Davis, B. Csuti, R. Noss, B. Butterfield, C. Groves, H. Anderson, S. Caicco, F. D'Erchia, T. C. Edwards, Jr., J. Ulliman, and R. G. Wright. 1993. Gap analysis: a geographical approach to protection of biological diversity. Wildlife Monograph 123.

Chapter 8

Employment in Aquaculture

ANITA M. KELLY

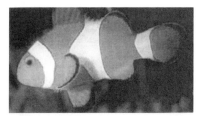

Aquaculture is the farming of aquatic organisms, including fish, mollusks, crustaceans, and aquatic plants, with some sort of intervention in the rearing process to enhance production, such as regular stocking, feeding, or protection from predators (FAO 2004). Aquaculture has been described as the aquatic version of agriculture; farming also implies individual or corporate ownership of the stock being cultivated. However, if you think that getting a degree focused on aquaculture means that you can only be a farmer the rest of your life, you are mistaken. While farming, agricultural science, and economics

are integral parts of aquaculture, the industry as a whole is dynamic, based on applied science (biology and chemistry), technology, practical skills, and experience.

There is a wide range of aquaculture industries, involving different aquatic plants and animals from abalone to zooplankton, and different technologies, such as tanks, ponds, and net pens. Aquaculturists not only work on farms, but they are also employed by aquaria, zoos, the ornamental fish (pet) industry, fisheries interests, marketing firms, and state and federal governments for policy development, research, and education. Many of these jobs

involve a variety of activities both outdoor and indoor, including field, laboratory, or office work, and working with organisms of all types and sizes in saltwater, freshwater, warmwater, and coldwater environments.

Globally, aquaculture plays an increasing role in the production of seafood. Currently, seafood imports are the second largest contributor to the U.S. trade deficit (Meyers 2005). Recent statistics show that approximately 38% of all seafood consumed globally is produced through aquaculture (FAO 2004). Aquaculture is also closely tied to fisheries management. Many wild fish stocks are supplemented with fish produced in private, state, and federal fish hatcheries. Millions of fish are stocked annually to support recreational fisheries, and fishing pressure on public waters is expected to double by 2030.

Aquaculture represents one of the fastest growing food production sectors, providing a product that is an acceptable supplement and substitute to wild catches. As we advance through the 21st century, aquatic food products will be in increasingly short supply as domestic and international demand for both high and low value species increases due to growing populations and increases in standards of living and disposable incomes. As global yields from traditional saltwater and freshwater fisheries remain static, requirements for aquatic products will largely need to be met by aquaculture. According to statistics compiled by the Food and Agriculture Organization of the United Nations, the contribution of aquaculture to global supplies of fish, crustaceans, and mollusks continues to grow, increasing from 6.1% of total production in 1979 to 31.8% in 2003 (Figure 8.1).

Aquaculture continues to grow more rapidly than all other animal meat sectors. Worldwide, the sector has grown at an average rate of 8.9% per year since 1970, compared with only 1.2% for capture fisheries and 2.8% for terrestrial meat production systems over the same period. Production from aquaculture has greatly outpaced population growth, with the world average per capita supply from aquaculture increasing from 0.7 kg in 1970 to 6.4 kg in 2002, representing an average annual growth rate of 7.2%. For food fish, over a quarter of total world supply is derived from aquaculture. Consequently, the future for employment in aquaculture and its related fields looks very promising.

Numerous job opportunities exist in the aquaculture field, and the skills and training required by the various jobs cover a wide range.

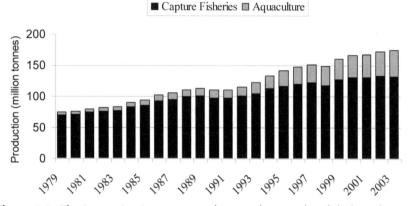

Figure 8.1. The increasing importance of aquaculture to the global production of fish, crustaceans, and mollusks.

Some jobs are fit for recent high school graduates while others require individuals with highly specialized advanced degrees. This chapter presents some of the job opportunities in aquaculture, as well as the skills and training required for successfully pursuing these careers.

Commercial Producers

Fish farming operations can be family-owned or private businesses, or they can be part of a corporation. The size of fish farms varies, and these facilities can engage in a single phase of production or they can participate in all areas of production. Consequently, depending on the type of facility, they employ individuals with different levels of formal training and experience. For example, some producers just produce fingerlings to sell to other producers to stock for grow-out or stock enhancement in various water bodies. These production facilities would require technicians with spawning and hatchery experience (Box 8.1). Producers that purchase the fingerlings and stock them into production systems for food size organisms would not require employees with spawning and hatchery experience; rather, they would need individuals with experience raising aquatic organisms from small to large sizes. Generally, the family or private farms are run by individuals

Box 8.1. Aquaculture Technician

Education Level: Two-year technical or community college degree

Responsibilities may include:

· breed and raise marine and freshwater organisms such as fish, eels, crustaceans, shellfish, algae, crocodiles, alligators, and turtles
· identify common diseases in fish and shellfish and take the necessary steps to prevent problems
· assist with experiments on nutrition or methods to control predators, parasites, and other disease-causing organisms
· monitor the environment using dissolved oxygen meters, salinity meters, pH (acidity) meters, and water chemistry analysis kits
· maintain live feed and algae cultures
· keep records of breeding, production, and treatment programs
· develop and implement systems of profitable farm management

Work Conditions

Aquaculture technicians work outdoors in all kinds of weather as well as in the hatchery or laboratory. Sometimes the work is located in isolated areas. Shift work is common, including on weekends and holidays.

Employment Sector

Commercial farms, government agencies, private organizations, and universities

who have hands-on experience from working on the farm when they were younger or have completed an Associate's, Bachelor's, or graduate (Master's or Doctoral) degree in aquaculture or fisheries. Larger corporations either have trained aquaculturists managing the farm or have a consultant that provides the necessary information for successful farming.

There are a variety of duties involved in farming aquatic organisms, ranging from spawning fish and raising live food to monitoring water chemistry and maintaining pumps. On the smaller farms, employees are usually skilled in all areas, whereas on larger farms employees have specific duties for which they are responsible. For example, certain employees are trained to monitor water quality (e.g., dissolved oxygen, ammonia, pH, nitrites) using either chemical test kits or electronic equipment, whereas other employees are responsible for detecting fish diseases, which requires routine monitoring of fish health using microscopes to examine fish and microbiological techniques to identify fish pathogens. The larger farms may have personnel responsible for culturing phytoplankton or zooplankton as food for larval fish. Having personnel on the farm capable of culturing these smaller aquatic food organisms may be essential. Fish farms must also have individuals that are responsible for maintaining pumps and other equipment. Some fish farms have their own crews to harvest and grade aquatic species for re-stocking or to haul them to processors for sale. Graduates of 2-year technical or community colleges or students with a Bachelor's degree are often sought as employees by commercial fish farms.

Larger farms and those owned by corporations may hire individuals with graduate degrees to conduct research and development. These individuals can be responsible for a variety of tasks, ranging from everyday farm chores to developing better husbandry practices or vaccines. Job opportunities exist for individuals with undergraduate and graduate degrees; however, the number of jobs available is inversely proportional to the level of training.

Government Agencies

The majority of the earliest aquaculturists in the United States were employed in government hatcheries to propagate and stock fish and invertebrates. In fact, the earliest fisheries management was simply culturing and stocking fish for recreational purposes, a mandate that still exists. Today, many local, state, and federal agencies employ aquaculturists. Individuals with Associate's to Doctoral degrees are employed by public agencies to provide assistance to the industry through extension offices and diagnostic laboratories. For example, the U.S. Department of Agriculture (USDA) has fish diagnos-

Box 8.2. Fish Pathologist

Education Level: Graduate or veterinarian degree

Responsibilities may include:

·   tend to the well being of farmed stocks
·   take all preventive measures to protect the farmed stocks from disease
·   administer proper treatments and medicines when necessary
·   maintain regular and detailed records of the farm, including stocking densities, feeding rates, days of treatment, and mortality rates
·   watch for signs of a disease outbreak

Work Conditions

Some fish pathologists may not work on the farm, but rather work for a government agency that provides diagnostic capabilities to area farmers. Those that work on the farm work mainly in the laboratory, but routinely sample fish or shellfish from the production system to check for disease.

Employment Sector

Commercial farms, government agencies, universities, support services (private companies specializing in disease diagnosis)

tic laboratories located in various locations throughout the United States, where fish pathologists diagnose fish diseases for culturists (Box 8.2). State governments employ aquaculturists in their agencies that deal with management of fisheries or natural resources. These individuals are usually working at hatcheries or involved with fisheries management of lakes, streams, and reservoirs. Offices of the Environmental Protection Agency located within individual states will also hire aquaculturists, as they generally have the skills necessary to make decisions on water quality. Other aquaculturists have been employed by private companies to assist with financial planning, insurance, or marketing. Agencies that employ aquaculturists at the federal level include the U.S. Department of Agriculture, the U.S. Fish and Wildlife Service, the Biological Resources Discipline of the U.S. Geological Survey, the National Oceanic and Atmospheric Administration (NOAA), the National Marine Fisheries Service (NMFS/NOAA Fisheries), the Bureau of Indian Affairs, and the U.S. Department of Defense.

Research opportunities in laboratories of various government agencies provide challenging employment opportunities for individuals with a multitude of specialties related to aquaculture. Chemists, biologists, physiologists, endocrinologists, veterinarians, engineers, nutritionists, and numerous other specialists are employed in laboratories in support of aquaculture programs. Most government jobs require specialized training at the level of a Bachelor's degree, while others require a Master's or Doctoral degree.

Private Organizations

Aquaria, zoos, and marine parks also employ aquaculturists. While some of these entities still collect fish from the wild, more are switching to culturing the fish in an effort to save wild stocks. Aquaculturists can also be responsible for display design and construction, and educational programs for visitors and schools. Employment opportunities are available for individuals with a Bachelor's degree or graduate degree.

Engineering firms will also hire individuals that have a background in aquaculture (Box 8.3). As the global water supply is becoming more of a precious commodity, culturists must develop techniques that reuse water. Many water reuse systems are in use today, but there is plenty of room for improvement and innovation in the engineering area. Individuals with a Bachelor's degree in aquaculture or aquaculture engineering can find employment opportunities here.

Pharmaceutical and biotechnology companies also hire individuals with aquaculture expertise. These companies investigate ways to treat diseases, improve stocks, and develop drugs, hormones, herbicides, and anesthetics, while making sure that the food is safe for consumers. These companies usually employ individuals with graduate degrees.

Box 8.3. Aquaculture Engineer

Education Level: Bachelor's or graduate degree

Responsibilities may include:

· design of tanks, cages, and ponds (includes carrying out the calculations for water flow, the material resistance, and the adequacy to the species)
· design and installation of sea cages and sea platforms
· design, selection, or building of equipment involved in production systems
· maintenance, adaptation, and repairing of specific equipment
· design and setup of equipment for energy supply, water quality control, and security
· design and setup of specific equipment related to feed manufacturing, feed maintenance and durability, feed processing, and quality control
· design and maintenance of equipment for fish and shellfish processing (feeders, collecting machines, processing equipment, cooling or freezing systems, environmental control, specific hatchery devices or installations)

Work Conditions

Most engineers work in the office drawing up designs and then work in the field testing the designs to make sure they function as needed.

Employment Sector

Support industries, commercial farms

Universities

Many universities employ faculty and research personnel with knowledge and expertise in aquaculture. Aquaculture programs are usually within departments of aquaculture, agriculture, biology, zoology, engineering, fisheries, or natural resources. Job opportunities range from administrative positions and professors to technicians and graduate students. Professors usually provide classroom instruction, conduct research both in the field and in the laboratory, and generally provide some extension services. Graduate students take classes, conduct re-

Box 8.4. Researcher

Education Level: Graduate degree

Responsibilities may include:
· study marine and freshwater organisms important to and potentially important to aquaculture
· evaluate the adaptability and conditions for the culture of specific organisms
· determine the genetic makeup of cultured species
· evaluate reproduction and ways to control reproduction
· determine nutritional requirements, effectiveness of different dietary sources, and their metabolic implications
· measure growth rates of aquatic organisms under different environmental conditions
· conduct research on pathologies and pathogens of aquatic organisms, their immunological traits, and disease prevention
· study statistics for the consumption of aquaculture products and public preferences
· research the economics involved in aquaculture production

Work Conditions

Aquaculture researchers work outdoors and in the hatchery or laboratory. The laboratories may be on land or onboard a vessel.

Employment Sector

Commercial farms, government agencies, private organizations, universities

search, and publish their work. Researchers are usually hired to conduct, analyze, and publish research findings (Box 8.4). Any individual conducting research can do studies dealing with the reproduction, nutrition, growth, and health of aquatic species. Individuals may study water conditions in relation to their effect on the species concerned. The researcher may be concerned with finding the best technologies for growth and maintenance of the aquatic species, or with studies concerning consumer preferences or future trends and economic projections for the sector or for a local industry.

The majority of job opportunities in universities are for individuals with graduate degrees. However, opportunities for individuals with a Master's or Bachelor's degree exist in the area of extension. The USDA has extension offices in each state which are usually associated with the state's land-grant university. Extension agents can provide information to persons wanting to get involved in aquaculture as well as people already in established businesses.

Support Services

A wide variety of job opportunities for individuals at all levels of training exist in support of the aquaculture industry. Individuals with training ranging from 2-year Associate's degrees through Doctoral degrees in fisheries or aquaculture often are employed to assist in developing and marketing supplies and services for the aquaculture industry. Hardware manufacturers and vendors employ individuals with Bachelor's and Master's degrees to supply tanks, aquariums, pumps, seines, 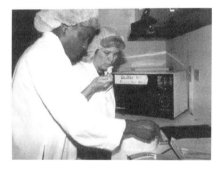 nets, analytical equipment, blowers, buildings, pond liners, aerators, and other equipment and supplies needed by the industry.

Consulting firms hire aquaculturists to assist in design and operation of facilities, to solve specific problems in a production facility, and to provide timely guidance to owners and investors of aquaculture firms. Fish processing plants, wholesalers, and marketing and advertising firms hire individuals knowledgeable of the aquaculture industry. Every fish feed manufacturer employs at least one individual with training in aquaculture. Some feed manufacturers employ aquaculturists to conduct in-house research and to provide support within the industry through extension activities. Some aquaculturists have moved into the retail business and work in pet shops to sell ornamental fish, aquatic plants, and supplies. Others are employed, principally in California and Florida, to inspect, package, or repackage live fish for interstate or international shipment.

Aquaculture Education Programs

Many community colleges and universities, both in the United States and elsewhere throughout the world, offer programs specializing in aquaculture. Such programs are particularly prevalent in the United States in the Pacific Northwest. The easiest way to find these programs is to conduct a simple Internet search for university or community college aquaculture programs.

Finding an Aquaculture Job

There are several places on the Internet where employment opportunities in aquaculture are posted. Some of the more popular sites include Aquanic, which has job postings for the World Aquaculture Society (http://www.aquanic.org), The European Aquaculture Society (http://www.easonline.org), the American Fisheries Society (http://www.fisheries.org), and the National Shellfisheries Association (http://www.shellfish.org). Other job listings can be found by conducting an Internet search.

References

FAO (Food and Agriculture Organization of the United Nations). 2004. The state of world fisheries and aquaculture 2004. FAO Fisheries Department, Rome, Italy.

Meyers, J. 2005. U.S. trade deficit statistics. Available: http://www.texasaquaculture.org/id156.html. (August 2005).

Chapter 9

Fisheries Employment in Canada

STEVEN J. COOKE AND
SCOTT G. HINCH

Canada has the second largest land mass of any country in the world and is surrounded by three different oceans (Atlantic, Pacific, and Arctic), with more than 265,500 km of coastline. In addition, Canada has more than 25% of the world's fresh water, much of it in the Great Lakes and the St. Lawrence watershed of central and eastern Canada. Large amounts of fresh water are also dispersed among several large watersheds in western and northern Canada (e.g., Fraser, MacKenzie, Nelson, Yukon). Canada has more than 2 million freshwater lakes spread across the country, covering about 7.5% of its land mass. Not surprisingly, the waters in and around Canada are home to an abundant and diverse assemblage of fishes. There are more than 23,000 registered commercial fishing vessels in Canada and more than 3.6 million recreational anglers. Of equal significance to the economies that these fisheries support is the role that fish and aquatic resources play in tourism.

There are diverse opportunities for fisheries-related employment in Canada in fields as disparate as pathology, limnology, anthropology, history, conflict resolution, restoration ecology, law, veterinary science, and, of course, resource management. The theme of diversity is one that will be emphasized in this chapter, consistent with Canada's diverse landscape, peoples, fisheries, and fisheries issues. We will attempt to extend this chapter beyond simply fish biology and fisheries management to include the many other professional fisheries positions available. Students should also consult other sources where appropriate. For example, a student interested in fisheries economics should also investigate employment guides that are general to the field of economics. The traditional pattern for students in fisheries science has been to obtain a degree in a biological discipline and then to work for a provincial or federal fisheries management agency. Presently, this pattern is rather atypical. Fisheries employment opportunities are now more prevalent in private industry and nongovernmental organizations than they are in government agencies.

Throughout this chapter we will attempt to emphasize issues that are unique or specific to fisheries employment in Canada. This chapter should supplement the more detailed accounts of different employment sectors covered in other chapters. We first discuss training and education required for different types of positions. The core of the chapter focuses on the current status of employment opportunities in key employment sectors. We conclude by providing insight into where to search for fisheries employment opportunities in Canada. We cannot provide names and contact information for all potential employers and we use Internet references sparingly, as they tend to become outdated quickly (some primary web sites are given in Box 9.1). We advise applicants to use common Internet search engines to look for specific opportunities after consulting this chapter as a starting point.

Training and Education

Almost all fisheries employment in Canada, especially at the professional level, requires some level of specialized training and education. Community colleges can prepare students for technician level positions with the provincial and federal governments. Increasingly, college graduates are assuming roles in fish culture settings. Some community colleges have recently transitioned into university colleges, offering Bachelor's degrees with more hands-on training than would be obtained through a typical university program. Bachelor's level education at accredited universities in Canada is probably the most common starting point for those seeking professional fisheries employment in Canada. In the past, students could take "fish and wild-

Box 9.1. Internet references for fisheries employment information specific to Canada.

Canadian Aquatic Resources Section of AFS	http://www.fisheries.org/cars
Canadian Association of University Teachers	http://www.academicwork.ca
Canadian Conference for Fisheries Research	http://www.phys.ocean.dal.ca/ccffr
Canadian Federal Government (Public Service Commission)	http://www.jobs-emplois.gc.ca
Canadian Society for Ecology and Evolution	http://www.ecoevo.ca
Canadian Society of Zoologists	http://www.uqar.uquebec.ca/jpellerin/csz
CareerOwl	http://www.careerowl.ca
Fisheries and Oceans Canada	http://www-hr.pac.dfo-mpo.gc.ca/pages/career_e.htm
List of environmental nongovernmental organizations, industry representatives, and other organizations	http://www.dfo-mpo.gc.ca/communic/ statistics/oceans/industries/index_e.htm
Monster.ca	http://monster.ca
Ontario Chapter of AFS	http://www.afs-oc.org
Society of Canadian Limnologists	http://uregina.ca/~scl
University Affairs	http://www.universityaffairs.ca/careers
Workopolis	http://www.workopolis.com

life" degree programs at Canadian universities, but this is increasingly rare in recognition that undergraduate students need to be provided with a broad background and skills that are applicable to more than just fisheries issues. Common degree programs suitable for individuals seeking fisheries employment include biology, zoology, ecology, marine science, integrative biology, animal science, environmental science, environmental studies, oceanography, resource management, or conservation biology. Some university programs provide opportunities to specialize or minor in areas such as aquatic biology, marine ecology, fish and wildlife biology, and aquaculture. Almost every university in Canada employs at least one fisheries or fish biology professor. Students can usually tailor their own degree to include aquatic-related courses, and can take field courses, complete internships or cooperative placements, or do a thesis to obtain additional hands-on experience. Employers are typically interested in finding individuals with both basic and applied training.

Increasingly, undergraduate students with interests in fisheries science are completing graduate level training. This provides an opportunity to further specialize and develop additional skills in research, problem solving, critical thinking, and communication that are both desired and essential for today's fisheries professional. For most, a Master's degree is sufficient, but there are also opportunities for Ph.D. level fisheries scientists in all sectors. For those interested in pursuing senior research careers (e.g., fisheries professor, research scientist), a Ph.D. and additional postdoctoral training are required. There are a number of training opportunities outside of degree programs where students and professionals can seek specialized training in topics such as electrofishing safety, habitat classification, statistical analysis, fisheries modeling, and biotelemetry. These courses are offered through colleges, universities, and professional societies such as the American Fisheries Society (AFS) or the Canadian Society of Environmental Biologists. Professional development and training is considered essential for maintaining professional certification (as through AFS; see Chapter 15) and is consistent with an interest in life-long learning. Also, with continuing developments in the fisheries field, it is essential for professionals to keep up with modern techniques so that they can be more effective in their current position and more competitive if seeking new employment.

Professional certification through AFS is not required by any of the federal or provincial natural resource agencies in Canada (unlike the

United States), nor do any universities develop their curricula such that it addresses all of the certification requirements. However, if interested in certification, a student can tailor his or her degree program accordingly and think more broadly about how to fulfill the certification criteria. For example, it is possible to use peer-reviewed publications on a topic in lieu of actually taking some course(s). Because the current AFS certification process is tailored to the U.S. educational system, the application process is challenging and those interested should consult the AFS Board of Professional Certification for additional guidance (e.g., allocation of course credits). Some jurisdictions in Canada do require a broader form of professional certification. For example, in British Columbia, the Registered Professional Biologist (RPBio) certification is required by most consulting firms and for some provincial government positions. This program is administered by the College of Applied Biology (CAB); the CAB Act is the first legislation of its kind in North America, and is the first time applied biologists have been granted full professional status through self-governing legislation. These certification requirements are typically met with comprehensive honors undergraduate programs in natural resources or biology to obtain "in training" status. Continued professional development is required to obtain "full" status.

Employment Opportunities in Fisheries

Academia

Canadian universities and some colleges are playing an increasingly important role in Canadian fisheries. Professors and their undergraduate and graduate students conduct applied research activities once conducted solely by government biologists. From a student's perspective, this provides many opportunities to obtain highly relevant experience while pursuing their degrees. Also, because much of this work is conducted in partnerships with agencies, students have an opportunity to collaborate with potential employers and get their foot in the door for employment following graduation. Many graduate programs require that graduate students be provided with research assistant positions as a means of providing partial financial support. Some programs and professors provide enough support through these research assistantships that additional support (e.g., teaching assistantships, bursaries, scholarships) is unnecessary. However, competition for graduate assistantships can be fierce and not all programs provide them. Students considering graduate school should carefully review the financial obligations of the professor and program before applying for positions. If students are able to provide their own funding through scholarship or contract support, then the student usually will have greater flexibility in choosing where and for how long they wish to study.

Technicians are usually hired during summer months to assist with academic research programs. These positions are often filled by senior undergraduate students. Research conducted by such technicians can develop into honors theses, provide training for potential graduate studies, or simply provide work experience for future jobs. Some of the larger fisheries laboratories have onsite wet laboratory facilities or field stations with dedicated professional staff to provide care to captive fish or to oversee field programs. These positions may be funded from departmental or program funds, which often ensures that these positions are longer term than that of most graduate or undergraduate research assistantships.

Postdoctoral or research associate positions are also considered employment (as well as training) opportunities, although most in the fisheries field are supported by some form of fellowship. For those interested in both research and teaching, becoming a professor is an option. There are relatively few professor positions advertised in Canada in any given year with the word "fish" or "fisheries" in the position description. This differs from the United States where there are frequent hires focused on fisheries science. In Canada, look for positions with emphasis on topics such as ecology or natural resource management and then think about how you can use fish as a model for that specific topic. Despite the fact that a large amount of the academic fisheries research in Canada is of an applied nature, the actual hiring of professors in this subject area relies importantly on that person's ability to conduct basic or pure science. This is a key element for receiving grant funding through Canada's largest granting agency, the Natural Sciences and Engineering Research Council of Canada, which funds research programs rather than research projects.

Local and Regional Government

Local and regional governments are increasingly becoming involved with fisheries issues and have growing opportunities for fisheries employment. Typically these positions are referred to as environmental specialists or water resource specialists and deal with day-to-day planning associated with development activities. Although these positions extend beyond fisheries, they do tend to have a mainly aquatic focus. Watershed-based management has also led to another layer of government. In Ontario, Conservation Authorities, which are primarily focused on water resource engineering, are conducting fisheries and aquatic science work including resource inventories and stream restoration. There are far too many regional governments for us to list—our message is simply that most local and regional governments are taking greater interest in aquatic ecology and opportunities abound.

Provincial and Territorial Governments

All provincial and territorial governments in Canada have a natural resources agency responsible for management and protection of fish and fisheries (note that some provinces instead have water resource agencies that include several individuals focused on fisheries). These agencies are typically tasked with natural resource management (including policy) and protection of biodiversity. The most obvious fisheries staff is conservation officers who deal with enforcement and outreach. These positions require specialized training in law enforcement and usually require a university or college degree in fish and wildlife or a related discipline. These positions are extremely popular and successful candidates usually have set themselves apart from the rest of the applicants through additional experience or expertise needed for working in remote locations (e.g., airplane license, dog handling, conflict resolution, small engine repair). Once hired into a pool of conservation officers, the trainees are exposed to additional in-house training in coordination with provincial police colleges.

Another common position is that of regional fish biologist. In reality, these positions are not specific to fisheries and often entail working on all sorts of resource issues, including plan reviews, stormwater management, recreational fisheries management, wetland protection,

stream restoration, and outreach. Most provincial agencies also maintain a group of dedicated fish biologists who deal with specific issues
(e.g., sportfish biologist, stock assessment biologist, science and technology transfer) or specific environments or water bodies (e.g., large
lakes biologist, stream ecologist). These positions are often more cen-
tralized rather than in district offices or at specific fisheries research
laboratories. Most new hires for biologist positions with the provincial
government agencies have Master's degrees. Also, most permanent
full-time hires are not in response to open job competitions. Biologists
must often accept a number of temporary government contracts before
having the opportunity to compete for internally advertised positions.
Contract employees often have some connection through earlier summer work experience. Blind application to these positions is rarely
successful; applicants should visit the office out of which the position
would be based.

The size of the agencies and their capabilities vary extensively
among provinces. British Columbia and Ontario both maintain a fisheries research branch staffed with Ph.D. level research scientists. Consequently, opportunities for fisheries research employment are highest
in these two provinces. Alberta and Quebec also have active research
programs. Large provinces such as British Columbia also employ fish
biologists in their provincial forestry department.

Federal Government

The Canadian federal government has a dedicated fisheries department—Fisheries and Oceans Canada (previously Department of
Fisheries and Oceans, or DFO; it is still widely referred to as DFO).
The mandate of DFO is to "develop and implement policies and programs in support of Canada's economic, ecological and scientific interests in oceans and inland waters." Fisheries and Oceans Canada
employs approximately 10,000 people across the country. During its
early years, DFO was regarded as the premier federal government fisheries agency in the world, with some of the most prominent scientists
producing scientific discoveries that formed the basis for today's modern discipline (e.g., William E. Ricker, Peter A. Larkin, Roly Brett).
However, financial cutbacks have reduced DFO's capacity for in-house
scientific research. There is still an active research component cover-

ing numerous topics (the DFO science branch is broken into Aquaculture, Environment, Fisheries, and Hydrology), but few of these positions are being refilled after individuals retire. Fisheries and Oceans Canada is actively partnering with academics and consultants to conduct much of their science. Thus, there are few opportunities for employment at the federal level as a research scientist. There are junior positions such as research biologists and technicians, but they are also difficult to get. Typically these positions are located at regional science centers (e.g., Pacific Biological Station in Nanaimo, Canadian Centre for Inland Waters in Burlington, Freshwater Institute in Winnipeg, Maurice Lamontagne Institute, St. Andrews Biological Station).

In addition to the research branch, DFO maintains a large policy and habitat protection group that focuses on conservation and sustainable resource use and protection of ocean environments and fish habitat. This group is also responsible for administration of the Fisheries Act, the primary piece of federal fisheries legislation in Canada that protects fish and fish habitat. For many years, administration and enforcement of the Fisheries Act was delegated to provincial government authorities. However, in the mid-1990s, DFO reassumed control over the Fisheries Act and consequently hired several hundred fish habitat biologists. Federal fish habitat biologists are located across the country; in Ottawa (headquarters), at regional offices in British Columbia, Manitoba, Ontario, Quebec, Newfoundland, and Nova Scotia, as well as in area and district offices within the regions. These positions have been largely filled with Bachelor's and Master's level biologists with general training in the biological sciences, rather than those with fisheries-specific training. Experience in restoration, mitigation, and monitoring are essential for these positions. Currently, this is still one of the best opportunities for recent graduates seeking employment with DFO. Similar to provincial agencies, the federal government also maintains an active role in enforcement and employs individuals having similar education and specialized law enforcement training as conservation officers. Fisheries officers deal with habitat infractions as well as enforcement of fishing regulations in federally-regulated fisheries.

Fisheries and Oceans Canada also has active stock assessment and enhancement units in addition to core fisheries managers. These positions tend to be regionally focused, dealing with specific species or fisheries (e.g., sockeye salmon *Oncorhynchus nerka*, Atlantic cod *Gadus morhua*, groundfish, Arctic marine and inland fishes). These units also have potential opportunities but are centered on the Pacific and Atlantic coasts. Much of the stock assessment and enhancement activities have been contracted to First Nations and consultants so there are fewer in-house positions than in the past. Fisheries and Oceans Canada also has opportunities in areas such as fisheries oceanography (e.g., Institute for Ocean Sciences in Sidney, Bedford Institute in Dartmouth), fish health and pathology, genetics, physiology, and administration. These more specialized fields can be further investigated on DFO web sites.

Other federal agencies also hire fisheries professionals, including Environment Canada, Department of Defense, and Agriculture Canada. The implementation of the Species at Risk Act has meant new biologist jobs in fish and wildlife conservation in Canada. Three departments administer this act; Environment Canada and Parks Canada deal with most wildlife species, whereas DFO has responsibility for fishes and shellfish in both marine and fresh waters. All of these agencies have been filling new positions in recent years. The Species at Risk Act calls for the assessment, protection, and restoration of organisms and habitats at risk. Much work is currently being done to develop an inventory of species at risk and to conduct species-specific status reports. Some of this work is contracted to organismal experts (e.g., academics, consultants), but much of the administration and policy work is conduced by Environment Canada employees. Here, training in conservation science and policy is a valuable asset.

Positions with federal agencies are advertised on the Internet through the Public Service Commission. Note that there may be geographic restrictions, such that applicants must have permanent addresses within the region of potential employment, but there are a number of positions with no geographic restrictions. For employment opportunities with the federal government, familiarity with the French language is an asset and in some cases is a prerequisite (fluency is required for some Ottawa-based positions). Detailed information on the requirements and necessary skills for each advertised position are provided in the online advertisement.

Transboundary Institutions

There are also some other government agencies that are relevant to Canadian fisheries but extend to other jurisdictions. The most prominent example is the Great Lakes Fishery Commission. The Great Lakes Fishery Commission was established in 1955 by the Canadian/U.S. Convention on Great Lakes Fisheries. The commission "coordinates fisheries research, controls the invasive sea lamprey, and facilitates cooperative fishery management among the state, provincial, tribal, and federal management agencies." Other such transboundary organizations exist in Canada, but tend to focus on a group of species rather than a specific geographic location. For example, on the Pacific coast, the Pacific Salmon Commission (PSC) was formed in 1985 after the Pacific Salmon Treaty was signed by the United States and Canada. The PSC focuses its efforts on the "conservation of Pacific Salmon in order to achieve optimum production, as well as dividing the harvests so that each country reaps the benefits of its investment in salmon management." Another group in the Pacific Northwest is the International Pacific Halibut Commission (IPHC). The IHPC was "established in 1923 by a Convention between the governments of Canada and the United States with the mandate to research and manage the stocks of Pacific halibut within the Convention waters of both nations." There are other commissions that deal with even more widely ranging animals that extend beyond just North America. For example, the International Commission for the Conservation of Atlantic Tunas (ICCAT; formed in 1966) focuses its efforts on "maintaining the population of tunas and tuna-like species found in the Atlantic Ocean and the adjacent seas at levels that will permit the maximum sustainable catch for food and other purposes." All of these and other agencies have some presence in Canada. If interested in one of these specific fisheries or locations, it is advisable to consult their web sites.

Consultancy

With government downsizing, much of the fisheries-related assessment or inventory work is being outsourced to consultants. The consulting field can be financially lucrative and somewhat stable, though work days can be very long. Some larger consulting companies hire full-time fisheries biologists but may also subcontract work to individuals, providing fisheries biologists with short-term employment.

Consultants may be hired to help develop plans that are compatible with government regulations. Often this work involves site visits, interactions with government biologists, and some level of report writing. Smaller firms or individuals working on their own tend to focus on an area of expertise and then collaborate on team projects with other consultants. There are also a number of very large Canadian firms with contracts and offices around the world. Depending upon the specific firm, one could either be focusing on a single project spanning multiple years or orchestrating a number of smaller projects simultaneously.

Typically consulting firms hire individuals with undergraduate degrees, but increasingly prefer Master's level training. In both cases, some regions may require some level of professional certification (e.g., RPBio in British Columbia). Some consulting firms will support employees to pursue graduate degrees part time while working for them. Many of the larger consulting firms also have a small complement of senior scientists with doctoral degrees. Often these individuals are also involved in the management and ownership of the company. Consulting firms actively recruit recent graduates, and in some summers find it difficult to locate enough qualified personnel for their jobs— this is an excellent way to obtain relevant experience early in one's career. Because of the broad power of the federal Fisheries Act, there is definitely a need for many more broadly trained aquatic biologists with expertise in fish and fish habitat that are being filled by consultants. Some consultancy firms are also involved in more research activities rather than resource inventories and assessments. Research activities tend to be less common for consultants and tend to be focused on applied fisheries issues, such as the impacts of hydropower operations or mining.

Fisheries Sectors

There are four primary fisheries sectors in Canada: recreational, commercial, First Nations, and aquaculture. All of these sectors employ a large number of fisheries professionals. Many of these opportunities are in private industry or with user group associations. For example, the recreational fishery has a large group of stakeholders in fields such as tourism, retail, and manufacturing, and they hire fisheries professionals to represent them in industry associations, committees, or working groups. In fact, many recreational fishing guides in Canada have some form of formal fisheries training. This can be a fantastic way to obtain early fisheries experience and interact with many stakeholders. In the commercial fishing sector, there are many gear technologists, advocates, observers, fishers, and product quality control specialists trained as fisheries professionals working to enhance the commercial fishing opportunities and efficiency. First Nations groups also employ a large number of biologists and technicians. Typically, funding for these positions is provided by the federal government and is ideally reinvested in those individuals in First Nations communities to build capacity (i.e., skills, knowledge) within their own community. The aquaculture sector also employs many people and is poised for future growth. British Columbia and New Brunswick produce about 80% of the cultured fish in Canada, with an estimated 5,000 to 6,000 personnel directly employed by the aquaculture industry. These positions range from general labor positions to husbandry professionals and those with advanced training in fish health. There are opportunities in this sector that also link to government and academia. The aquaculture industry in Canada has been directly supporting research into topics such as nutrition and epidemiology.

Other Opportunities

During the 1990s, environmental nongovernmental organizations (ENGOs) increased in Canada by 50% and in some jurisdictions (e.g., British Columbia) by over 300%. Over 2,000 persons are currently employed in Canada in this sector. This shift in employment patterns is clearly due to downsizing of government agency responsibilities and probably to increasing awareness and concern of environmental issues by the public. There are a large number of ENGOs that promote conservation of fisheries resources and employ fisheries professionals. There are a number of museums and aquaria in Canada that employ outreach, collections, maintenance, and research staff. There are also opportunities in the private sector with utility companies. For instance, hydropower companies typically employ fisheries professionals for planning, assessment, and research. Canada is also a leader in fishing gear development and scientific instrumentation. In fact, several of the world's leading fish telemetry firms are Canadian. There are opportunities for fisheries professionals in these and other organizations and industries—be creative in your searching! Fisheries and Oceans Canada maintains a *very* comprehensive online list of ENGOs, industry representatives, and other organizations with relevance to fisheries in Canada (see Box 9.1).

Where to Search for Advertisements for Canadian Fisheries Posistions

Locating graduate student positions in Canada does not differ substantially from that in the United States or elsewhere (see Chapter 3). Research assistantships are often posted on the web sites of the Canadian Society of Zoologists, the Canadian Society for Ecology and Evolution, the American Fisheries Society, the Society for Conservation Biology, the Ecological Society of America, and the American Society of Limnology and Oceanography. The Canadian Conference for Fisheries Research also maintains a listserv where Canadian student fisheries opportunities are posted. Summer employment or research assistantships for undergraduate students are rarely advertised formally. These are largely word-of-mouth positions and should be sought by direct contact with professors who have research programs that align with your interests.

For those interested in nonacademic positions, the same web sites mentioned above are appropriate. Individuals interested in working for provincial governments need to consult the individual agency or provincial web sites. Similar efforts are required for those interested in working for the federal government. In fact, all applications for positions with DFO and other federal departments are announced through the Public Service Commission. Unfortunately, government positions are rarely advertised for long periods, usually less than a month, and thus require constant monitoring to identify possible positions. Consultant positions can be advertised on web sites as above or on web sites of the individual company. Consulting firms and ENGOs also tend to visit environmental fairs, job fairs, or conferences with the sole purpose of recruiting. Because consultants often get large projects suddenly, students should do informational interviews in person with the personnel manager or relevant biologist at the company of interest. Having met you and having your resume in hand helps them find you if unexpected jobs come along.

Those interested in professorships or postdoctoral positions at Canadian universities should consult *University Affairs*, the online publication of the Association of Universities and Colleges of Canada. The Canadian Association of University Teachers also has an online publication called *Career Bulletin*. In addition, most Canadian academic positions are also posted to *Science* magazine or the *Chronicle of Higher Education*. Individual departments or university web sites are also worthy of a visit, as not all positions are advertised more broadly (usually due to timing limitations). This is especially the case for limited term and sessional positions that tend to materialize on short notice. Applicants should also consider contacting peers in potential departments to inquire about future hiring plans. Most academic hires are projected several years in advance, so one can gain insight into what positions may or may not be on the horizon.

As with all employment opportunities, knowing, or being known to, the employer or one of their team members can represent a significant advantage. Networking at conferences is one way to make such contacts. However, the conferences that you attend with job seeking in mind should be compatible with your target position. For example, attending the Annual AFS Meeting is not the ideal setting for obtaining a position in regional government, unless it is held very close to where you wish to work. Instead, job hunters would be better served by attending more regional or topical conferences such as AFS divisional or chapter meetings. Serving in a leadership role on a committee can also generate valuable contacts. Canadians interested in obtaining positions elsewhere (principally the United States) should also actively participate in international committees. For example, AFS attempts to obtain Canadian participation for all committees, often with little success. This is an important way to contribute to the fisheries profession, develop a broader network of colleagues, represent Canada, and improve one's resume or curriculum vitae.

It is also important to note that the North American Free Trade Agreement (NAFTA) has made it possible for students and fisheries professionals to obtain cross-border employment in either direction.

Although immigration issues are often complex, there are few positions where jobs are not open to the international community. Although federal regulations require verbiage in job descriptions about preference for citizens from their own country, in most cases the primary interest is in hiring the most qualified person. Academic institutions, governments, and private industry all have the ability to hire international applicants and will usually assist with the immigration process.

Chapter 10

International Fisheries Employment

RICHARD A. NEAL

Choosing a foreign assignment over a domestic one can be a decision that dramatically affects your career in fisheries science and your future. When the job, your personality, and other collateral factors like family all mesh together well, foreign assignments can be immensely rewarding and provide you with a pathway towards international prominence in fisheries science. On the other hand, foreign assignments have greater potential to present unexpected problems than do jobs in your homeland. This chapter provides guidance about the nature of foreign positions in fisheries science and introduces potential avenues of international employment and sources of information.

Advantages of a Foreign Assignment

The advantages of a foreign position can be substantial and range in nature from philosophical to financial. Working abroad carries with it an element of adventure and almost inevitably brings you into contact with animals and ecological systems that are different from those in your home country. Indeed, the elements of adventure and "newness" may be the primary reasons young professionals seek international employment opportunities. Major advantages of foreign employment are intellectual stimulation and the broadening of your knowledge about other cultures and different fisheries. Insights and perspectives obtained while working abroad can help you grow personally and make you a better fisheries scientist for the remainder of your career. In many cases, these aspects of your experience in a foreign assignment can make you more competitive for subsequent domestic employment. If your ambition is to achieve an international reputation in fisheries science, there is hardly a better vehicle than assignments abroad. They allow you to meet and interact with peers from many nations. Your opportunities for international exchange are typically much greater if you hold an overseas position.

Another reason for accepting a foreign assignment is that, for one or a combination of reasons, the remuneration can be better than that of available domestic positions. Such reasons can include higher salary than that for equivalent domestic work, substantial perquisites, lower cost of living, and tax advantages pertaining to money earned out of the home country. Often the basic pay is greater with an international agency, a foreign university, or a private firm operating abroad than the pay for equivalent domestic work. Employees of the U.S. government, when holding assignments in international agencies to which the United States is party, usually receive remuneration from either the international agency or the U.S. government. The remuneration includes not only net pay, but also living allowance and an educational allowance for dependents.

The fringe benefits and perquisites attached to foreign positions are often substantial as well. A common fringe benefit is annual paid travel for the employee, and sometimes dependents, to and from the home country. Vacation time is usually ample and housing may be provided. United States government employees temporarily attached to international agencies can retain the health and life insurance benefits of their employing agencies, as well as re-employment rights, retirement credit, and other benefits.

A third benefit accruing to scientists working abroad may be lower living costs. Especially in developing countries, the costs of food, housing, and domestic services may be lower than at home. Families living in developing countries may achieve a lifestyle available only to the wealthy in the United States and Canada. On the other hand, living expenses in some countries are higher than they are in the United States and Canada. A prudent prospective employee will make a careful examination of costs likely to be encountered when working abroad.

Canadian and U.S. citizens employed abroad by foreign governments, universities, and private businesses (but, in the case of U.S. citizens, not by the U.S. government) may be eligible for substantial tax benefits if the condition of nonresidency is established in the home country. Currently, U.S. citizens may exempt up to $80,000 in income from U.S. tax liability if they were out of the United States for an entire tax year, or for 330 full days in any consecutive 12-month period. Nonresidency may be harder to acquire for Canadians. Because tax rules are volatile, anyone considering foreign employment should check policies at the time of employment. Income acquired abroad may also be subject to taxes in the country of residency.

A fifth benefit of working abroad is the opportunity to travel in and around the country of employment. Additionally, flights to and from the country of employment are often available with alternative routes with free or inexpensive stop-over privileges. Thus, global travel subsidized by the employer may be possible.

Disadvantages of a Foreign Assignment

The old principle of "no free lunch" applies to foreign fisheries employment just as it does to ecology. The substantial benefits of working abroad may go hand-in-hand with adversities that, depending on the employee, range from amusing challenges to insurmountable difficulties. Perhaps the most significant contributors to culture shock are psychological, philosophical, and otherwise abstract aspects of life in a foreign country. Prior to accepting employment, a candidate should investigate conditions in the country of future residency to the greatest extent possible. Here I review a few of the most prominent issues likely to be encountered in an overseas assignment.

A major contributor to culture shock can be the necessity to function in a second language. The need for language skills is usually clearly defined in employment announcements, and most foreign workers possess sufficient knowledge of the required language to function professionally (co-workers may often be fluent in English). However, if the language skills of the visiting professional do not allow him or her to function fluently on the street, in social gatherings, or in the marketplace, a sense of isolation and alienation can develop. On the other hand, there is no finer training in a foreign language than full immersion by necessity.

A second major source of adversity peculiar to some foreign positions is increased risk to physical health and welfare. For example, a host of diseases not found in North America occur in some tropical countries, and special preparations or exceptional diligence may be required to avoid problems. Cholera, malaria, plague, and yellow fever still exist in some parts of the world. By and large, however, the worst illnesses are prevented by required inoculations or prophylactic drugs. Common sense will prevent others. Air pollution levels in the larger cities of the developing world may also be of concern, especially to persons with respiratory problems. Potable water is not available in many public water systems, but safe bottled water is nearly always available. Standards of sanitation for storage and sale of meats and produce and disposal of sewage and waste differ from country to country. As in the United States, crime rates can be much higher in some areas than others. In all remote areas, especially of developing nations, medical care may not meet the standards of North American practice. However, modern medical facilities are probably available within a few hours flight of most airports.

A common source of disillusionment is the nature of government bureaucracies that seem even more inscrutable, multilayered, and counterproductive than those in the home country. Whereas the structure of government may fit well with the expectations and attitudes of the permanent inhabitants, its functioning may appear arcane or even nonexistent to the North American mind. Given that fisheries professionals often are intimately involved with the local government, patience, diplomacy, flexibility, and adaptability are required. Although some overseas governments are extremely effective, a valuable cliché is "look before you leap," as a poor choice could result in major frustration.

In addition to government structure, cultural attitudes in the host country can greatly affect your success and happiness. Perhaps the most important attitudes to consider are those relative to work and work ethics, women in the workplace, alcohol and drugs, and gender relationships. Whereas zeal for labor in foreign countries may equal or exceed that in Canada or the United States, industriousness as a cultural attribute is not universal. Attitudes about work intensity, work hours, and absenteeism may not be what the visiting North American expects. Completing projects and meeting deadlines may not be as important to native professionals as they are to the visiting investigator.

The position of women in the professional workforce in North America has advanced markedly, but such advances have not occurred worldwide. Cultures exist abroad wherein women will be at a distinct disadvantage when working as a fisheries professional. Even when women working abroad find relative equality and respect among native male coworkers, these same women may not enjoy freedom of movement and expression among the general populace.

Attitudes about the use of alcohol vary widely among cultures and could create pitfalls for a foreign investigator. For example, in some Muslim countries the use of alcohol is forbidden even to foreigners and visiting investigators risk social ostracism or legal action if they indulge in even modest use of alcoholic beverages. Similarly, recreational use of some psychotropic drugs is tolerated in some parts of the United States, but foreign visitors must be especially heedful of local laws about drugs. Use or possession of even minor amounts of drugs in other countries could result in long imprisonment under primitive conditions, or worse.

Not all cultures share the relatively liberal European–American attitude about sex and gender relationships. Adultery still warrants death in some places. On the other hand, some cultures are much more relaxed about sexual relationships. Depending on the host country, visiting professionals may encounter situations for which they are unprepared and should exercise due discretion and caution.

Logistical difficulties associated with the availability of certain resources in the host country may also affect research activities and personal comfort. For example, electricity is absent or only sporadically available in many rural regions of the world. In addition, whereas research instruments, computers, outboard motors, and vehicles may be available, repair services may be poor. Lastly, whereas means of communication with the rest of the world have generally improved in recent decades, they may be neither fast nor reliable in some areas.

An important factor in the response of any worker to conditions overseas is his or her family situation. In some positions and areas, an individual who has a family with him or her is somewhat insulated from life in an unfamiliar culture and may be happier than a single person in responding to the difficulties of a new lifestyle. Conversely, some assignments may entail environmental demands of a nature that would engender concern about the health and safety of a family, especially small children. Provision of high quality education for children may be difficult.

Of course, few or none of the potential difficulties I have listed may occur at any particular position. I provide the listing to emphasize that resources that are taken for granted in North America may not be available elsewhere. Even when and where they do occur, difficulties need not detract from the enjoyment of a foreign assignment. If the changes in lifestyle and facilities are viewed as challenges, overcoming them will add richness to your life and memories of the experience.

Types of Positions Available

Although some research positions are available overseas in fisheries science, openings are more likely to be related directly to fisheries management, collection of data related to management, aquaculture production, and teaching. For aquaculture, technology continues to advance with regard to genetics, reproduction, nutrition, and diseases, and leaders in these specialty areas may be in considerable demand around the world. As trade in fisheries products increases and hygienic standards receive more public attention, industrial expertise in the handling, packaging, and preservation of fisheries products is being sought by commercial firms as well as some governmental agencies and universities. The development and marketing of new specialty fisheries products is a related area attracting interest.

Statisticians, computer programmers, stock assessment experts, and modelers are widely in demand, particularly in developing countries. Personal computers are now widely available, even in developing countries, but training in full use of these computer capabilities is needed. Application of available software, programming to meet specific needs, and interpretation of the results of packaged software applications are areas where assistance will be useful. Statisticians who can provide leadership in experimental design and development of fishery sampling protocols may also have opportunities for employment abroad.

In all countries, businessmen and politicians tend to favor short-term resource exploitation over longer-term sustainable uses. Fisheries managers need supporting personnel who can help them justify and promote appropriate management measures to ensure sustainable, economically sound use of resources. Specialists needed to assist management include stock assessment biologists, fisheries economists, and increasingly, sociologists. The special role of economists in working to limit harvesting capacity has not yet been applied in most countries, and, particularly in developing countries, social concerns such as employment often override efforts to improve economic efficiency of the industry. Furthermore, large-scale environmental concerns are growing around the world, but fisheries expertise is not often brought into efforts to solve environmental problems. Skills in organizing and addressing the political, social, economic, and biological aspects of large environmental issues are in demand.

Some fisheries administrators with a depth of experience in organizing and implementing complex programs are hired for foreign positions. Individuals with a proven track record of successful administration of fisheries research or management organizations can expect to be competitive for these positions. Prior experience in foreign countries is an asset in these assignments.

Special Qualifications

For any foreign assignment, several special personal qualifications are important. Besides an adventurous spirit, persons interested in foreign assignments must have flexibility—a capacity to promote ideas vigorously without taking rigid stances; adaptability—the ability to adjust easily to changes in working environment, information bases, and cultural values and expectations; good health—the ability to handle additional stress, air pollution, and gastrointestinal disturbances and illness; and compatibility—the ability to work closely with people who have a different view of things.

Potential Employers

For various political, economic, and practical reasons, almost every employer would prefer to hire adequately trained local personnel rather than hire foreigners. Therefore, anyone interested in working outside the United States and Canada should have something special to offer in the way of training, experience, or capabilities. Exceptions to this are the use of unskilled or minimally skilled volunteers by the U.S. Peace Corps or other voluntary organizations that are involved in low-technology fishing or aquaculture assistance to impoverished communities. Such opportunities can be especially rewarding and, for recent college graduates, overseas volunteer work in fisheries adds significantly to their list of experiences and qualifications.

An experienced fisheries scientist with many years of experience and a clear area of specialization may find numerous opportunities for foreign employment, particularly if the area of specialization represents new applications of technology. In developing countries, where many employment opportunities exist, the skill level of the applicant must match the level required by the position. Various positions for both experienced and less experienced individuals may exist in these countries. The following are brief discussions of the primary potential employers for fisheries professionals interested in a foreign position.

International Organizations

International organizations established to promote cooperation, development, or rational resource use are the principal sources of foreign opportunities for employment. The Food and Agriculture Organization of the United Nations (FAO), for example, has an active fisheries program of broad scope that provides development assistance. Other groups may stress research (e.g., the WorldFish Center), environmental concerns (e.g., United Nations Environment Programme), management of shared resources (e.g., the Secretariat of the Pacific Community), or development loans (e.g., Asian Development Bank). Experts in a wide variety of fisheries and related fields may find interesting opportunities with these groups. Pay scales and benefits are designed to be attractive to experienced fisheries experts from all countries.

Foreign Governments

Foreign governments may require special expertise or exceptionally capable individuals for expert or advisory roles. These positions are often temporary or short term and are likely to include technology transfer and training activities. Positions are usually within the fisheries or natural resources department of the country. Poorer countries are less likely to have native professionals with all of the skills needed, particularly those relating to new technologies. The poorer countries also typically have low government pay scales, making positions with them less attractive. Some countries use international organizations or consulting firms through contractual or other arrangements to hire foreigners at international rates, thereby avoiding national limitations on pay scales.

Private Voluntary Organizations

These organizations serve a wide variety of purposes ranging from community development to resource protection and their organization is similarly diverse. These groups often promote development, and some include fisheries or aquaculture in their work (e.g., Aquaculture without Frontiers). Both those with considerable experience as well as the inexperienced may find opportunities with these groups. Pay is typically low for assignments in developing countries, but benefits such as food, room, and vehicle may compensate for low salaries. Work with a private voluntary organization is often a route to foreign work with other organizations as it provides experience in a foreign setting and an introduction to foreign cultures.

Consulting Firms

Some consulting firms specialize in providing expertise and special assistance to foreign governments, U.S. government agencies, and international organizations for both the short and long term. Those interested in hiring U.S. citizens have offices in the United States. Positions typically require substantial expertise and duties may range from evaluations or reviews lasting 2 or 3 weeks to multiyear assignments. Project planning exercises are also often conducted by teams hired by these companies. Consulting firms also conduct long-term development projects for agencies such as the U.S. Agency for International Development, The World Bank, and the Canadian International Development Agency. Foreign governments may solicit specialized expertise through consulting firms. Some consulting firms work with one or more U.S. universities to provide needed expertise from their faculties and staffs. Personnel to train extension workers, fishing vessel observers, aquaculturists, and various other professionals are in demand in some instances.

Private Companies

International commercial firms are active in foreign countries and may have openings for individuals with biological, fisheries, or aquaculture experience. Joint fishing ventures may require expertise with gear, vessel operation and maintenance, or handling and processing of products. For example, the shrimp culture industry in Latin America had ties with U.S. companies in its early days and hired hatchery, feed, and disease experts from the United States. The increasing international market for quality fish products has increased need for product handling experts.

Universities

Many universities have an associated "institute" designed to be a business arm of the university and to handle contracts, commercial activities, and funding from outside sources for research. These institutes often hold development contracts in foreign countries, particularly if the contract includes research, training, or education and matches the expertise of faculty members. The institutes may also join forces with consulting firms to help supply the expertise needed to fulfill a particular development contract. The institutes tend to look for expertise first among their faculty and students, but may also hire experts without ties to the university. Universities may also have strong ties with other countries or foreign universities with arrangements to exchange faculty, educate graduate students, or provide long-term assistance of an educational or research nature (the International Center for Aquaculture and Aquatic Environments, ICAAE, at Auburn University is an excellent example). Teaching of fisheries science at foreign universities is a frequent area of employment.

U.S. and Canadian Government Agencies

The U.S. Peace Corps has placed large numbers of volunteers with some basic aquaculture training in developing countries. These volunteers tend to work in community development to introduce the concepts of aquaculture. In addition, the U.S. Agency for International Development and the Canadian International Development Agency fund some fishery and aquaculture projects that are normally staffed through contracts with consulting firms or universities. The U.S. Department of State has fishery attaches in several overseas posts who serve as regional contacts for international negotiations on fisheries affairs and legal matters.

Finding and Preparing for a Foreign Assignment

Preparation for and selection of an overseas assignment requires substantial research. Word of mouth can be a valuable source of infor-

Box 10.1. Internet references for various entities offering opportunities for international employment or volunteer work.

Aquaculture without Frontiers	http://www.aquaculturewithoutfrontiers.org
Asian Development Bank	http://www.adb.org
Canadian International Development Agency	http://www.acdi-cida.gc.ca
Conservation International	http://www.conservation.org
Food and Agriculture Organization of the United Nations	http://www.fao.org
ICAAE, Auburn University	http://www.ag.auburn.edu/fish/icaae
Inter-American Tropical Tuna Commission	http://www.iattc.org
International Commission for the Conservation of Atlantic Tunas	http://www.iccat.es
International Council for the Exploration of the Sea	http://www.ices.dk
International Development Research Centre	http://www.idrc.ca
National Marine Fisheries Service/NOAA Fisheries, Office of Sustainable Fisheries, International Fisheries Division	http://www.nmfs.noaa.gov/sfa/international
Secretariat of the Pacific Community	http://www.spc.org.nc
Sierra Club	http://www.sierraclub.org
The Nature Conservancy	http://nature.org
The World Bank	http://www.worldbank.org
United Nations Environment Programme	http://www.unep.org
U.S. Agency for International Development	http://www.usaid.gov
United States Peace Corps	http://www.peacecorps.gov
WorldFish Center	http://www.worldfishcenter.org
World Wildlife Fund	http://www.worldwildlife.org

mation about job opportunities overseas, but young fisheries professionals are not likely to have a professional network of an order that provides access to such information. Other than word of mouth, information about foreign positions is rather diffuse and locating it may require detective work on your part. Fortunately, the Internet can be an invaluable tool in this task, providing access to information on employment opportunities (e.g., job announcements) as well as the culture, standards of living, and political climate in the location of the position (Box 10.1). Good sources for job announcements are the AFS Job Center Online (http://www.fisheries.org/html/jobs.shtml) and the related listings in *Fisheries* magazine. In addition, classified advertisements in *Science* and *The Economist* often list positions at foreign universities. Finally, the web site and members of the International Fisheries Section of AFS (http://www.fisheries.org/ifs) can be an important source of information. As a follow-up to internet research, direct contact with potential employers or volunteer organizations is always informative and is encouraged. If possible, try to interview other North Americans working in the area so that you can be prepared and know what to expect.

Suggestion for Additional Reading

Segal, N., and E. Kocher. 2003. International jobs: where they are, how to get them. Basic Books, New York.

Chapter 11

Fisheries Employment with Nongovernmental Organizations

WILLIAM E. PINE, III AND
KENNETH M. LEBER

A nongovernmental organization (NGO) is typically noncommercial, nonprofit, and independent of government inputs or control. Just as the functions of many governmental fisheries management agencies have expanded beyond their traditional roles, such as managing stocks to meet consumptive demands, the role of NGOs in natural resource management has grown as well. Many NGOs are conducting original fisheries research, assisting with policy development to protect and enhance aquatic ecosystems, and helping to organize stakeholder groups with common interests. Given that the number of active NGOs is increasing and their role in natural resource management is growing, NGOs are important employers of fisheries professionals and they should not be overlooked.

Nongovernmental organizations exist for a variety of reasons, but they are primarily focused on furthering the common goals of their members. The World Bank defines NGOs as "private organizations that pursue activities to relieve suffering, promote the interest of the poor, protect the environment, provide basic social services, or undertake community development." Nongovernmental organizations are typically classified as being either operational or advocacy-oriented. Operational NGOs are primarily concerned with the design and implementation of development-related projects, whereas advocacy NGOs are interested in defending or promoting a specific cause. This simple dichotomy is generally accurate, but many NGOs involved in fisheries science and management spend nearly equal amounts of time working in both areas. For example, Trout Unlimited (TU) devotes about half of its resources to designing and implementing on-the-ground restoration projects. However, TU is also involved in advocacy, as stated in the mission of the organization—"to conserve, protect and restore North America's trout and salmon fisheries and their watersheds." Advocacy NGOs can also include smaller organizations that are primarily concerned with local issues, such as invasive aquatic plant management on lakes or reservoirs.

Education is an important component of the mission of nearly all NGOs. Most organizations actively engage in programs designed to educate their membership, the public, and legislators and other government officials about their mission. Educational activities may also be directed specifically at younger people (K–12) in the local community, or at broader audiences via the Internet, newsletters, and magazines.

Professionals are usually attracted to NGOs because they are dedicated to the specific purposes and principles of the organization. In addition, NGOs are often much smaller than government employers, and the close working relationships among employees, opportunities for travel, casual atmosphere, and schedule flexibility appeal to many professionals. Employees of NGOs typically deal with many of the same challenges and opportunities encountered by public or private sector fisheries professionals. However, because of the independent nature of NGOs, their employees often have access to more outlets for research, development, and advocacy than those afforded by other employers. For example, a biologist working for a NGO may be able to develop international components of their research more easily than a government biologist because of fewer jurisdictional concerns. Nongovernmental organizations are also often involved in advocacy for specific species or issues, whereas government agency biologists may be prohibited from taking on advocacy roles. Nongovernmental organizations can provide unique opportunities for fisheries professionals to affect decision making about natural resources by linking science, policy, and advocacy.

Educational Requirements

Most college and university curricula for fisheries science and management emphasize biology, ecology, taxonomy, fisheries management, and statistics, while requiring few courses in areas such as human dimensions, socioeconomics, public relations, and public speaking. Although some schools are tailoring courses in social sciences and human dimensions specifically for students in fisheries and natural resources, such courses are not prevalent in most curricula. Nongovernmental organization employees will be most successful if they have a balanced education in all of these areas. In addition, courses in accounting, political science, environmental policy and law, journalism (nonscientific writing), and education can be valuable for students who are considering employment with a NGO. Furthermore, because NGOs are often small, unique skills such as web site design, fluency in foreign languages, computer management and troubleshooting, photography and graphic arts, or fundraising may be key to securing employment with these organizations. As a result of the breadth of education and experience that NGO employees need (Box 11.1), many NGOs hire professionals that have graduate degrees in addition to Bachelor's degrees (see Chapter 3).

Employment with a NGO can be a good fit for anyone that is dedicated to the mission of the organization. Such individuals could include people without education or experience in fisheries who are changing careers, perhaps from business, law, engineering, or education, into fisheries science, management, and policy. Individuals with these and other skill sets may be able to apply their skills in a unique way to help the NGO accomplish its goals, while learning more about fisheries.

Types of Positions Available

The jobs available to fisheries professionals through NGOs will vary tremendously depending on the mission and size of the organization. Some positions may be similar to those available in the academic, government, and private sectors (see Chapters 4–7 and 12), but

Box 11.1. An example of a fisheries biologist position with a NGO.

Education and Experience

Bachelor's degree required; Master's degree preferred. Must have
strong organizational and interpersonal skills, an ability to work
constructively with individuals from a variety of partners, outstanding
problem-solving abilities, and excellent written and oral
communications skills. Prior experience working with citizens groups
and the media or developing outreach and education activities is
helpful.

Responsibilities may include:

- Developing, initiating, and managing fish habitat restoration and
  population assessment projects
- Coordinating and conducting outreach activities for landowners,
  governmental groups, and local and regional media
- Assisting staff with fundraising efforts, including identifying
  potential sources of funding, drafting grant proposals, and
  developing relationships with donors

Work Conditions

Diverse work conditions depending on the task at hand. May include
field sampling activities in all weather conditions as well as
professional meetings with partners and government agencies.

NGO jobs will have additional expectations related to the mission of
the organization. A key difference between NGO positions and those
in other areas of fisheries is that NGO employees are often actively
engaged in advocacy in order to further the goals of the organization.
The advocacy aspect of working for an NGO affects employees at all
levels and makes employment with NGOs distinct from more traditional fisheries employment opportunities (e.g., state or federal agencies).

Nongovernmental organizations are usually involved with a diversity of issues and projects. As a result, most organizations provide
employment opportunities for professionals at a variety of career stages.
In this section we present some general classifications of employment
opportunities with NGOs. Nongovernmental organizations usually have
a combination of advocacy, education, research, and fundraising sec-

tors, and each employee is typically expected to contribute to some degree in every area. In particular, a significant portion of each employee's time may be spent on grant proposals and other fundraising projects.

Biologists, Managers, and Researchers

Most NGOs are deeply involved in the hands-on management of land and water resources through acquisitions of land, procurement of easements, discussions of property rights, or research projects. Some NGOs purchase tracts of land to be held perpetually in trust or to be transferred to public agencies (e.g., The Nature Conservancy). Such areas are often unique ecologically, and employment with these NGOs can offer opportunities to identify, study, and facilitate acquisition of valuable natural areas. Nongovernmental organizations often hire biologists to study the resources on acquired lands and to work with government, private landowners, and other NGOs to develop effective management plans.

Some NGOs are increasing their involvement in original fisheries research. Such involvement may include a NGO partnering with a university or agency to conduct cooperative research related to the NGO's mission (e.g., habitat restoration), or a NGO providing funding, technical, or logistical assistance to the individuals conducting the research. In either case, a NGO will typically have a biologist on staff who is actively involved in the research project to ensure its success and report back to the organization on how the research is contributing to the NGO's mission. Several large NGOs are even more strongly focused on both basic and applied fisheries research. For example, Mote Marine Laboratory has evolved from a small independent research laboratory into a large nonprofit organization with seven applied research centers. In addition, Mote has a substantial internship program through which students can participate in research projects and potentially earn course credit at their academic institution. At Mote and similar large NGOs (e.g., Harbor Branch Oceanographic Institution, Hubbs-SeaWorld Research Institute, Ocean Institute) volunteer and employment opportunities are available in both research and education (typically focused on K–12 and continuing education), and some operate large aquaria or other educational facilities. These facilities are often designed to generate operating expenses as well as funds to support ongoing research.

Issue Analysts

Many NGOs work directly with government agencies to develop and implement management programs that benefit the resources of concern to the organization. For example, the Snook Foundation is primarily concerned with management issues that affect common snook *Centropomus undecimalis* populations in Florida, such as stock replenishment strategies, harvest regulations, and habitat loss. This foundation also supports research aimed at filling information gaps, thus enabling resource agencies to adopt adaptive management strategies and conduct better resource assessments. When presenting information to policy makers about a particular issue (e.g., a change in harvest allowances), it is extremely important that the testimony of NGO employees be credible. Therefore, an issue analyst for a NGO must be a capable and respected scientist with a solid grasp of the technical issues, but he or she must also be able to clearly communicate these issues and the NGO's position to policy makers. Consequently, employment opportunities as issue analysts are most often available for "senior" fisheries professionals—individuals with many years of experience, a broad network of professional contacts, and an understanding of both the biological aspects of an issue as well as the management principles and processes that are involved. Individuals with a broad education in resource management and diverse experience working with government agencies will be most successful in these positions.

Government Relations

Government relations programs are an important activity of any NGO, and involve NGO representatives interacting with policy makers and agency personnel to ensure government support for actions and regulations that help achieve the goals of the NGO. Experienced fisheries professionals commonly enter into government relations positions with NGOs after having worked their way through the administrative ranks of state or federal management agencies. In addition, professionals who have served in government positions, especially those with a political or legal focus, may develop an interest in a particular NGO and be able to serve as a government liaison. Individuals involved with government relations must be flexible, able to work well as part of a team (e.g., with representatives from other NGOs), able to work on a variety of issues of concern to the NGO, and thoroughly knowledgeable about legislative and regulatory processes.

Working for a NGO

Similar to private consulting firms (see Chapter 12), employment with a NGO will probably be less secure than other types of fisheries employment, such as government agency jobs. In addition, NGOs are not generally as restricted by hiring procedures as government agencies, and are thus able to hire individuals rapidly when needed, even if an individual does not have all of the formal training called for in the position. Initial employment opportunities with NGOs are often based on short-term funding and are thus limited to the time frame of a specific project. However, if an employee excels and actively engages with other professionals involved with the project, these initial opportunities can lead to a network of other employers and perhaps more permanent employment.

The number of fisheries professionals employed by a NGO will depend on the mission and size of the organization. For example, a NGO like Trout Unlimited will have a much greater proportion of its workforce dedicated to fisheries than an organization with a broader resource conservation focus, such as The Nature Conservancy. Large NGOs with national and international operations may employ hundreds of individuals, but most of them will probably work on things other than fisheries issues. In addition, large NGOs may be able to afford to hire more people and narrow down employee responsibilities, whereas employees of smaller organizations may need to be involved in a wider array of tasks.

The salaries for many positions with NGOs are partially funded by "soft" money—funds that are not permanently in place. In these cases, the employee is expected to support a portion of his or her salary through research grants or fund-raising activities. Having your salary dependent on your success as a researcher, grant-writer, and fundraiser can be highly motivating and can create salary incentives (e.g., grant-generated raises) that are not available to public agency employees.

Volunteering and Internships

A great way to gain professional experience and learn more about working for a NGO is by volunteering or through an internship. Many NGOs rely on volunteers and interns to help them accomplish their goals, and opportunities are usually available year-round. Because NGOs tend to be relatively small and lack complex bureaucracy, interns can often be involved in a number of issues or projects that are outside of their area of formal training or expertise. These opportunities are particularly valuable for identifying areas of interest or disinterest and for developing diverse work experiences, some of which may mature into completely new talents. Internships may be paid or unpaid, and sometimes offer other incentives such as housing for work in remote or unique areas. Volunteering and internships provide young professionals with opportunities to showcase their talents, and may give them an inside track on future job openings with a NGO or elsewhere. Individuals who are already employed but are contemplating a position with a NGO should consider volunteering for the organization to help them decide if such a position is right for them.

Personal Visits

When seeking employment with a NGO, personal contacts established by visiting an office and meeting with the staff is critical. A personal visit will make a good impression with the organization's leaders and demonstrate that you are truly interested in their mission and the work that they do. In addition, you can get a feel for the attitude and approach of the organization by meeting and talking with current employees.

Conclusion

Employment with NGOs can be extremely challenging and rewarding. The opportunities available for fisheries professionals through

NGOs are diverse and unlike opportunities available through more traditional avenues. Nongovernmental organizations can provide opportunities for individuals with a variety of skills who may not fit into a traditional agency position, and can offer that "new opportunity" that many professionals look for at some point during their career. Nongovernmental organizations are playing an increasing role in the study and management of natural resources, and fisheries professionals should consider the opportunities offered by these organizations.

Chapter 12

Private Consulting in Fisheries Science

DONALD MACDONALD, FORREST OLSON, AND ANDREW J. LOFTUS

Depending on your perspective, employment opportunities provided by consulting firms represent either the chance of a lifetime or the bane of your existence. Regardless of your point of view, however, private consulting firms are providing an expanding share of the fisheries employment market and should be considered by serious job hunters. Recent surveys indicate that a substantial percentage of professional fisheries biologists are employed by consulting firms (e.g., 14%; AFS 2004).

Small consulting firms typically employ one to ten people and are often highly specialized. As a result, these types of firms tend to target relatively small and focused projects. However, by developing cooperative associations with other independent consultants, other small firms, or larger firms, small firms can acquire and leverage the expertise necessary to bid on bigger multidisciplinary projects. In addition to fisheries scientists, medium and large firms often employ a wide variety of professionals, including hydrologists, geomorphologists, geologists, engineers, economists, social scientists, environmental policy analysts, and planners. Ready access to a range of skills enables larger firms to take on large, multidisciplinary projects. As such, fisheries biologists in larger firms have an opportunity to work on a variety of projects, of which fisheries issues may be only one component. Some firms specialize in bioengineering projects, such as land and water management, hatchery design and fish passage, and fisheries and environmental policy, while others focus on assessment, management, and restoration of fish populations and habitat. Projects are often conducted to address federal environmental laws, such as the Clean Water Act, the Endangered Species Act, the National Environmental Policy Act, and the Comprehensive Environmental Response, Compensation, and Liability Act. An understanding of these laws and their requirements is an important part of fisheries consulting and can lead to interesting and rewarding work in fisheries conservation and restoration.

Fisheries professionals have a variety of roles and responsibilities in the consulting field. The consulting biologist often plays the role of liaison between his client (often industry) and the agencies charged with protecting resources. In these cases, the consulting biologist may provide ongoing management of the project, focusing on problem assessment and remediation and providing management recommendations when things go unfavorably. In other cases, the consultant or firm works directly for the natural resources agency (i.e., federal, state, or provincial), providing skills that the agency needs but does not possess. This role can be especially rewarding to those interested in the "people" aspect of fisheries work rather than just the technical details. In recent years, watershed advocacy groups and other nongovernmental organizations have been increasingly seeking the services of consulting biologists to solve complex watershed planning issues by balancing the diverse interests and needs of multiple stakeholders. In this role, fisheries biologists must understand the social, economic, cultural, and logistical issues that influence the feasibility and acceptability of watershed plans.

Consulting firms provide a number of employment opportunities for students, recent graduates, and established professionals. Undergraduate students are attractive to independent consulting firms because a number of government and university programs are available to subsidize student wages. Cooperative education and employment programs encourage firms to hire students on a short-term basis. In return, the students gain experience that is invaluable to them in securing full-time employment after graduation. Many independent consulting companies also hire recent graduates on a part-time or temporary basis, thereby providing new professionals with excellent opportunities to gain practical experience. Recent graduates or established professionals can also develop their own small consulting firms, providing highly motivated individuals with an opportunity to excel as an entrepreneur.

Qualifications

Consulting firms tend to operate with tight deadlines and focused budgets. Therefore, employers are often looking for people who can contribute immediately. When evaluating undergraduates, employers generally look for good grades and some work experience, but also flexibility and eagerness to work. Strong computer skills are a must, and employers frequently seek individuals with advanced skills in database development, data management, statistical analysis, modeling, and geographic information systems. It is especially important to maintain an emphasis on data quality and sound science in applying these skills, being careful to avoid overextending study results and explicitly recognizing uncertainty. In addition to these general qualifications, individuals seeking employment with consulting firms should have a four-year college or university degree, relevant work experience, proven communications skills, and the ability to work on a team. Individuals who think of themselves first and the team second need not apply!

Consultants employ individuals with all levels of education (Bachelor's, Master's, and Doctorate), but a Master's degree will definitely be an advantage in getting a job. A degree specifically focused on fisheries should not be viewed as a necessity. Private firms are not tied to rigid hiring guidelines like those attached to most civil service jobs, and thus have the flexibility to hire anyone that possesses the qualifications necessary for a particular job. Because of the diversity of projects that consulting firms work on, a varied background in fisheries as well as other related fields is always attractive to a consulting company.

Whereas a college or university degree is important, relevant work experience is what counts most in getting hired by a consulting firm.

What qualifies as relevant experience can be quite varied though, depending on the type of clients and projects in which the consulting firm specializes. A large number of firms work on projects involving habitat alteration or restoration rather than traditional fisheries management. More and more consulting firms are becoming actively involved in ecosystem studies, watershed analyses, and cumulative impact assessments, so any experience or training in these areas would be considered an asset when applying for a job. Additionally, some nonprofit organizations and companies require consultants that understand politics and know how to navigate the fisheries management arena. Building and expanding unique skill sets is a great way to distinguish yourself from other candidates and provide a future employer with a product that can be marketed to prospective clients.

One of the best ways to get the work experience required to gain employment with a consulting firm is to work for government agencies, industry, or academic institutions as an undergraduate or early in your career. Alternatively, volunteering in community-based habitat enhancement or rehabilitation projects often provides important experience and demonstrates initiative and devotion. Employers look for individuals who are serious about their profession. Show your commitment by attending professional meetings, preparing presentations, publishing research papers, and obtaining professional certification (see Chapter 15). In addition, try to obtain specialized certifications to increase your areas of expertise (e.g., wetland delineation). All of these activities will increase your confidence, develop and hone your skills, and make you more attractive to potential employers.

Communications skills are always high on the list of qualifications sought by consulting firms. The primary work products of consultants are reports and even the best scientist will not be an effective consultant without the ability to convey information in writing. It is important to work on consistently improving your writing skills and we recommend that you consult tested references to help you become more proficient (e.g., CBE Style Manual Committee 1994; Strunk and White 2000; Fowler and Aaron 2003). Lastly, remember that plagiarism is a serious offense, so be sure to give proper credit when citing publications or data of other scientists.

Landing a Consulting Job

The job market in the field of fisheries and aquatic resource management is very competitive and is becoming more so every year. To land a job, you must stand out from the rest of the crowd. The first step is to prepare an impeccable resume or CV and a convincing cover letter (see Chapter 2). These items are the potential employer's first introduction to you as a professional and provide you with an opportunity to demonstrate your written communication skills. Be sure to highlight your relevant work experience and volunteer efforts, but be careful not to overstate your qualifications or desire for the position. If you are hired for a job that you cannot adequately complete, it will not be a satisfactory move for you or your employer. Credibility and integrity are vital and you will find that a well-earned reputation will provide dividends throughout your career.

Consulting firms often do not prepare and distribute advertisements for a position. If they are looking for someone with special experience or capabilities, they usually first spread notice of the position opening among professional contacts in agencies, universities, other firms, or professional organizations. For entry-level positions, they often hire students who worked for them during a summer and stood out or rely on their own files of recently submitted unsolicited resumes. Therefore, it is important for you to contact firms frequently so that they think of you when they are looking to hire someone. Make an effort to visit the office to get more information about the firm and its activities and offer information about yourself. Demonstrate a sincere willingness to work hard. In many ways, this approach is superior to waiting to be called for an interview because it is informal and neither party feels pressured into making a decision. In addition, you will gain important information to help you decide if you really want to work for the firm. For more senior positions, consulting firms often recruit people from government agencies or other firms who they have worked with on collaborative projects in order to get the right person for the job.

Whether you schedule an informal discussion with someone in the company or are called for a formal interview, it is essential that you make a good first impression. It is useful to treat the interview like an exam—study for it. Make yourself familiar with recent studies in the field, especially those with which the firm may be involved. Inquire about regional fisheries and environmental programs and issues. It is this type of effort that will set you apart from other applicants. Do not be surprised if you are offered short-term work or a position related to a specific project at first. Private firms rarely make a long-term commitment without knowing that an individual can make a significant contribution to their team. Even if you are not offered longer-term employment with the firm, this experience will prepare you for employment elsewhere.

Advantages of Working for a Consulting Firm

There are several advantages to working for a consulting firm. First, if you demonstrate a willingness to accept responsibility and have the capabilities needed to participate effectively on a team, you will excel. In addition, this type of employment offers a variety of work experiences, limitless opportunities to learn new things, and some ability to select projects of interest to you. Moreover, you will have an opportunity to work on current, relevant issues and make a real difference in the management of resources. Over the long term, working for a private firm or developing your own company will also provide you with a means to succeed financially. If your ultimate goal is to start your own consulting firm, working first for a successful firm will provide you with the skills and experience you will need to strike out on your own. You will gain a firsthand understanding of the day-to-day operations of running a business, as well as all aspects of project management.

Disadvantages of Working for a Consulting Firm

Perhaps the biggest downside to working for independent consulting firms is the lack of job security when compared to government or academic jobs. While business has generally been good for consulting firms over the past decade or so, there are no guarantees beyond the contracts that are already signed. In addition, salaries are typically low over the first few years, although they generally improve substantially as an individual becomes established in the field. Overtime work is usually compensated on a hour-for-hour basis, so do not expect to augment your salary with weekend and holiday work. Importantly, retirement, medical, and dental plans are often not provided by small firms or are less extensive than you may desire, and these items can put an added strain on your finances.

Consulting companies often have a "work hard, play hard" philosophy regarding vacation time. Your schedule will frequently be driven by tight deadlines to meet a client's needs. At times, unexpected issues related to projects can destroy the best-laid vacation plans. However, if you can live with some uncertainty and pressure, employment with consulting firms can provide challenges and job satisfaction that are unparalleled in other sectors.

References

AFS (American Fisheries Society). 2004. The strategic plan of the American Fisheries Society, 2005–2009. Prepared by the AFS Strategic Plan Revision Committee (2003–2004). American Fisheries Society, Bethesda, Maryland. Available: http://www.fisheries.org/html/fisheries/governance/afsplan.shtml. (April 2006).

CBE (Council of Biology Editors) Style Manual Committee. 1994. Scientific style and format: the CBE manual for authors, editors, and publishers, 6th edition. Cambridge University Press, New York.

Fowler, H. R., and J. E. Aaron. 2003. The little, brown handbook, 9th edition. Longman, New York.

Strunk, W., Jr., and E. B. White. 2000. The elements of style, 4th edition. Longman, New York.

Chapter 13

Advancing to a Career in Fisheries Administration

STEVE L. MCMULLIN AND CHRISTOPHER HUNTER

Few people entered the fisheries profession because they wanted to work with people and fewer still aspired to a job that would place them behind a desk. In fisheries, as in many technical fields, the reward for excelling as a biologist is usually a promotion into a position that requires less use of biological skills and more use of managerial, leadership, and administrative (i.e., people) skills (Figure 13.1). For that reason and others, many fisheries professionals show little interest in moving up the ladder within their organizations (McMullin 2005). They view administration as the dirty work of dealing with personnel issues, politics and budgets. So, why should young fisheries professionals consider administration as a viable option in their career paths? In this chapter, we discuss some reasons why fisheries professionals should aspire to be administrators, as well as some considerations for preparing for an administrative career. Although our discussion focuses on administration in state fisheries and wildlife agencies, the principles apply to administrative roles in many organizations.

Among the many good reasons to consider administration as a career option, is that for many fisheries professionals, long days in the field lose some of their allure as they age and gain experience. As their careers progress, the different challenges and rewards of administration in fisheries may gain more appeal. In Box 13.1, we offer 10 good reasons why young fisheries professionals should consider administration as a career.

Box 13.1. Ten good reasons why young fisheries professionals should consider administration as a career.

1. <u>You will earn more money</u>—Few people enter the field of fisheries out of a desire to become wealthy. On the other hand, we all need to be able to make a comfortable living and provide for our families. In addition, if you make more money, you can buy more fishing gear and be better able to afford your kids' college expenses. Most retirement systems base your retirement check on the highest salary you earn over a period of time, often three years. It is not unreasonable to expect to collect retirement for 20 years or more, so the impact of a higher salary can be very important to a financially secure retirement.

2. <u>You will influence policy</u>—If you don't like the policy direction of your agency on issues ranging from resources to personnel, you can either gripe about it while doing your field work or you can become a policy maker. As an administrator, you will have great influence on agency policy.

3. <u>You can improve morale</u>—Administrators have the ability to affect the morale of their entire fisheries staff. Higher morale among the staff affects not only their work efforts but all aspects of their lives. Becoming an administrator provides an incredible opportunity to affect peoples' lives in a positive way.

4. <u>You can create change</u>—Change is difficult, but essential for individuals and agencies. Administrators often lead their organizations in making the changes that influence the success of an agency and its employees.

5. <u>You will gain recognition</u>—You will be respected for the title you carry. People recognize the title as an indication of achievement and respect you for that. Both authors have taken great pride in representing not only the current program and staff, but all those who went before us. It is a great honor and responsibility.

6. <u>You will only be in the field when you want to be</u>—At some point in your career, it is likely that those night electrofishing adventures in the cold and sleet will lose their appeal. As an administrator you decide when and where you want to be in the field. Staff members will be sure to take good care of you when you show up.

7. <u>You can shape an organization to your vision</u>—Do you have a vision of where you think your organization should be headed in the future? You will have an excellent opportunity to make that vision a reality in an administrative position.

8. <u>You have an opportunity to mentor the next generation of fisheries professionals</u>—As an administrator, you will have the opportunity to help younger biologists develop into the future leaders of your organization. For many administrators, this is the most rewarding and important aspect of their jobs.

9. <u>You find the human dimension of fisheries management exciting</u>—We often hear biologists say "we don't manage fish anymore, we manage people." We have always managed people and that aspect of the job can be incredibly interesting and exciting.

10. <u>Your career goals will change as your career progresses</u>—Today you may have great interest in the habitat requirements of a certain species or some other aspect of fisheries biology. As your career progresses you may find yourself increasingly interested in policy development, mentoring, or creating a vision for your organization. As you progress through your career, keep your mind open to the possibility of moving into administration to meet your personal career goals.

Administration is closely associated with positions of leadership in organizations and it is in the role of leader that administrators make their marks and draw much of their job satisfaction. The purpose of this chapter is to describe what fisheries administrators do and how to become an effective administrator. Much of our discussion will focus on aspects of program leadership. However, it is important to remember that all but the lowest level of positions in the fisheries profession include some administrative duties. Technicians in seasonal positions have few, if any, administrative duties. Nearly everyone else in the profession must supervise employees and participate in team-oriented activities that help to govern the organization, and these are administrative activities. As Figure 13.1 illustrates, the amount of administrative activity increases as you move up the organizational ladder, while the amount of time spent on technical aspects of fisheries conservation decreases.

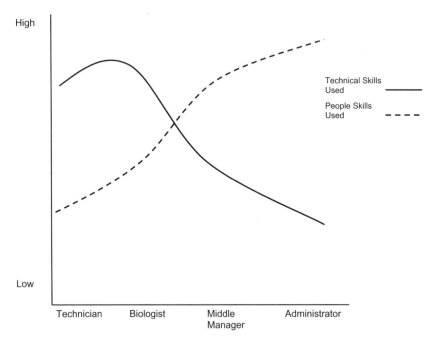

Figure 13.1. Relative amounts of technical skills and people skills used by technicians, biologists, middle managers, and administrators in the fisheries profession.

What Administrators Do

Develop Policy Direction for the Fisheries Program

The overall direction of a fisheries program is determined by policies that are established by politicians, agency directors and fisheries administrators. Well-articulated policy gives staff and the public a better understanding of the areas of emphasis of the fisheries program. Good policies also allow the field staff some latitude in their implementation of policies. Policy issues may deal with fishery resources and their use, personnel, or fiscal issues. Some examples of policy decisions that could have far-reaching implications for a fisheries program include a policy to manage for wild fish rather than stocking hatchery-raised fish, a policy to standardize field methods among all regions of a state, a policy to incorporate nongame fish and amphibians in all survey efforts, a policy on how to deal with illegal introductions of fish in public waters, a policy to protect all genetically pure native fish populations, and a fish health policy designed to strictly limit the spread of fish diseases by governing movement of fish within the state and leaving state and private hatcheries.

Administrators ultimately are responsible for the development and implementation of the policies that provide direction for the fisheries program. By choosing the subjects of policy and by following through with implementation of the policy, an administrator can have a major impact on both the current and future condition of the fisheries resources of the state. Development of policy cannot be unilateral, as pronouncements from on high are not generally well received. The real challenge in policy development is to develop a sense of ownership and acceptance for the policy among the staff and the public.

Determine Priorities for the Fisheries Program

Determining the priorities for the fisheries program can be one of the most challenging aspects of the administrator's job. Difficult challenges facing the administrator include deciding among a number of things he or she might like to accomplish, such as giving underpaid and overworked employees a raise, bolstering their flagging operations budgets so they have the funds to purchase safe equipment, and starting a new program to restore stream habitat in the state. The job becomes even more demanding when budget cuts force the administrator to decide how to cut back programs. Although not a pleasant task, this is one of the places where an administrator has the opportunity to have a lasting impact on the program, doing what he or she feels is the best thing for the fisheries resources and the people who enjoy them. Determining priorities for the program provides an opportunity to build a team atmosphere and should reflect considerable thought and discussions with key staff members.

Allocate the Budgets to Implement the Program

Once policies providing overall direction for the fisheries program are in place, you and your management team can determine priorities for the program based on those policies. Then, you can allocate the budget needed to implement the program. No other responsibility of the administrator has a greater immediate impact on the fishery re-

source. That responsibility is tempered by the reality of budgetary
mandates that reduce an administrator's discretion in budget allocation.

You may find that things have been done in a certain way for a
number of years. Changing the way things have been done can be
threatening and unsettling and you can expect to meet with resistance
from those most directly affected by changes you intend to implement.
Development of policy and establishing priorities is, to some extent, a
paper exercise, but allocating budgets is real and is one of the primary
responsibilities of an administrator. Allocating budgets is difficult work
but it is what allows you to take the fisheries program in the direction
of your vision. As we discuss in more detail later, staff members watch
administrators closely to see if their actions match their words. This
includes watching to see if budget allocations match stated priorities.

Represent the Program before the Public, Commissioners, and Legislators

An administrator makes presentations to the agency's commission
(a group of citizens who have policy-making authority for most, but
not all, state fish and wildlife agencies) and the legislature that are
critical to advancing the fisheries program. It is important that administrators develop a good rapport and working relationship with the commission in order to advance the fisheries program. In federal agencies,
administrators may represent their programs in front of agency direc-
torates or in Congressional hearings. The commission ultimately makes
many of the big decisions that guide the program. They must approve
all fishing regulations and they approve land acquisitions designed to
improve access for fishing or for protection of habitat. Several surveys of Montana anglers have found that they think the two most important issues are access and habitat protection. If the administrator is
not successful in convincing the commission that land acquisitions for
access and habitat protection are important and affordable, he or she
fails the stakeholders. In many states, the commission also must approve management plans for species or specific waters.

The stakes are even higher with the legislature or Congress, as the
entire budget and key legislation hang in the balance. In the months

prior to a legislative session, the fisheries program management staff must develop proposals, budgetary and legislative, important to further the fisheries program. These proposals must be convincingly presented to the rest of the department's administrative staff in order to become part of the legislative package the department takes to the legislature. Once the proposals are part of the legislative package, the administrator must develop presentations that will be concise and compelling to legislators. He or she then may be called upon to present these proposals and answer some of the craziest questions imaginable. We never cease to be amazed and amused that the legislators (who, in the western United States, often hail from agricultural backgrounds) often approve multimillion dollar program requests with little apparent review, while spending substantial time discussing whether the agency is planning to purchase the right kind of tractor for a wildlife management area. The administrator also will prepare testimony to support legislation the fisheries program is pursuing, or to oppose legislation viewed as detrimental to the program.

As an administrator you also will be a key spokesperson for the fisheries program and for the entire department. You will be invited to attend and speak at banquets held by sportsman and conservation groups. You will serve as a dignitary at fishing access site openings. You will speak to service groups, school kids, and college students about the fisheries program. These speaking engagements can be a lot of fun and can go a long way toward furthering the goals of the program.

Participate in Workgroups Determining Agency Priorities and Direction

As a fisheries program administrator, you will be asked to work on agency committees that develop policy direction for the entire agency on a wide variety of topics. In recent years, the second author has participated on a department management team committee that has been trying to determine how to incorporate the State Wildlife Grant program (a Federal Aid in Fish and Wildlife Restoration program designed to support efforts by state agencies to conserve nongame fish and wildlife) into his agency. It has been interesting and challenging to increase attention paid to nongame fish and wildlife in a traditional fish and game agency.

These department-wide policy level committees deal with a wide variety of issues. Over the past year, the second author participated on a committee that discussed how to allocate a 4% pay increase approved by the legislature. We have grappled with questions such as, should everyone receive the same 4% increase or should raises be given on the basis of performance? Should administrative staff persons, who normally make a lower hourly wage, receive a higher percentage wage increase than the professional staff? The second author also has been involved with a technology committee that is struggling with a myriad of issues surrounding data management and technology development. How much of the sportsmen's dollars should we spend on web development? What data are we required to make accessible to the public? Should publicly available data include the location of nesting and spawning areas of threatened or endangered species? Serving on these department-wide committees provides the administrator with a broader understanding of the issues the agency is facing. You also learn from the experiences of other program administrators who probably have faced some of the same issues you are trying to work through in the fisheries program.

Participate in National Workgroups that Set National Policy and Goals

The American Fisheries Society and the Association of Fish and Wildlife Agencies are two organizations that work on national policy issues related to fisheries and wildlife. They both provide information to policymakers that is designed to improve and support conservation of fish and wildlife resources. These organizations provide many opportunities for administrators to participate on national committees that work on vital policy issues. For example, the State Wildlife Grant program, referred to earlier, provides funding to state fish and wildlife agencies to work on species in need of special management. These fish and wildlife species often are already listed under the Endangered Species Act (ESA) or otherwise are species of special concern. Currently the State Wildlife Grant program is funded by Congress on an annual basis. Although the additional funding is helpful to the agencies, it is difficult for them to develop programs when they are not certain that future funding will be available. The Association of Fish and Wildlife Agencies and other organizations are in the forefront of the effort to see this program funded permanently, similar to the Wallop-Breaux sport fish funding mechanism. If and when this program is funded on a permanent basis, it will fundamentally change the way that state fish and wildlife agencies go about their business. No longer will the work be so narrowly focused on game species. Native fish and wildlife (other than game species) that have been largely neglected for years will gain their own source of funding, a reflection of a growing constituency for native and nongame fish and wildlife in the country.

Participation on national committees is a great way to meet your peers from around the country. It is an opportunity to learn how they have dealt with some of the same issues you will probably face in your fisheries program. The contacts you make working on these committees also may lead to other opportunities, professional as well as recreational (e.g., fishing with the best guides in the world—your peers).

Hire, Mentor, and Supervise Key Personnel

Making personnel decisions is challenging and often stressful because you directly affect the lives of the people that work for you. While agonizing over a hiring decision, the first author once was advised by a senior official of the agency, "Every time you make a hiring decision, you change the lives of all the people you choose or reject forever." In the long term, making personnel decisions can be more important than allocating budgets. If you have made the budget allocations to implement program priorities, but you don't have the right personnel in place to implement the projects, your best-laid plans will fail. On the other hand, if you have done a good job of identifying the skills of your staff, put them in the positions where they can best use their skills, and have given them the tools to do their jobs, watching them work and succeed at implementing the priority work of the program is extremely gratifying. Because administrators, by definition, lead programs, they are judged by the success or failure of the program (and thus the accomplishments of staff) rather than their individual efforts.

Hiring new employees is one of the most important decisions facing supervisors. When you hire a new employee, you may be hiring the next chief of fisheries or director of the organization. Or you may be hiring someone that will be a constant source of trouble for you and the agency. We cannot overstate the importance of hiring the right people. When you hire a new employee, look for someone who has a positive attitude, has lots of energy, is bright, and will get along well with the rest of your staff. You also should consider how a prospective staff member's background complements those of existing staff. Of course, this all will be done within legal constraints designed to ensure fairness and diversity, which is still an important consideration in the largely white, male fisheries profession (see Chapter 14).

Mentoring staff occurs at several levels. Staff members will watch you to see how you deal with the commission, the legislature, staff, stakeholders, and difficult situations. Even when you are not consciously trying to teach someone, people are learning from your successes and failures. Mentoring also occurs at a more personal level when you work with an individual to help him or her learn what it takes to be successful in the organization. Many agencies have formal mentoring programs that provide some structure and coaching for the participants. However, effective mentoring occurred for many years before the advent of these programs. Nearly all of us benefited early in our careers from the guidance of one or two mentors. If you spot a young biologist that you feel has the right stuff to be a good administrator some day, let that person know that you are available to help him or her learn more about leadership and how to succeed in the organization. Few aspects of the job create more satisfaction than seeing a young biologist that you have mentored become a leader in the field.

Supervising key staff should be an enjoyable part of your job. You have hired good staff and you have provided them with the training and mentoring to succeed within the organization. Supervision should be as simple as providing them with the tools they need (budget, staff, and time), giving them meaningful work to do, providing them with clear direction and giving them the room to be creative and make decisions. You should be generous in your praise and rein folks in when necessary. You also should try to find creative ways to reward those staff members who do excellent work. Creativity on your part is important because often it is not possible to compensate an outstanding staff member financially due to the agency compensation plan.

Our combined experience suggests that people working in fisheries generally are very bright, hard working, highly motivated, and dedicated employees. In general, your biggest personnel challenges will be providing the staff with the tools and direction they need and putting them in the best position to take advantage of their strengths. Inevitably, however, a few employees will require closer supervision and perhaps, even disciplinary action. These are difficult situations that some administrators will avoid because they are not pleasant to address. However, these are among those situations in which you will be closely watched by your staff. Avoiding them not only hurts morale within the work place, but it does not help the employee and in the long run, makes dealing with the situation even more difficult. A good administrator will deal with these issues rather than letting them fester. Allowing an employee to take advantage of the agency can only foster ill feeling among the rest of the staff and it delays inevitable action at the expense of the agency, the public and the fishery resources. It is just as important, much easier and more fun, to recognize and acknowledge the outstanding work of most of your staff. You cannot do this enough.

Create the Work Atmosphere within the Division and the Agency

The administrator has an opportunity to create a productive work atmosphere for the entire staff. If the administrator controls information, does not delegate important work and decision making, is dictatorial, vindictive or moody, the workplace is unlikely to be productive or fun. The staff, fisheries resources, and the stakeholders all will suffer. If, on the other hand, the administrator is free with information, gives the staff important and meaningful work, the freedom to make decisions and to be creative in their approaches to problem

solving, is supportive, provides positive feedback, and is approach-
able, friendly, and even happy, the result is more likely to be a positive, productive atmosphere where a great deal can be accomplished.
The administrator unconsciously influences the work atmosphere of
the central staff and extending to the entire division, including the biologists and field technicians that he seldom has a chance to get to
know personally.

Set the Standard for Work Ethic, Scientific Rigor, and Tone for Dealing with Issues

Creating an atmosphere conducive to creativity and productivity is
an important aspect of managing a program, as is setting the expectations for how work will be accomplished. As we mentioned previously, the staff will learn a great deal about what your expectations are
by watching how you conduct your work and they may emulate your
approach. If you show a strong work ethic, respect for scientific rigor
and public opinion, and you generally are open and forthcoming with
the public and other agencies in dealing with issues, your staff will
likely take a similar approach. If you take a different approach, the
staff will notice that as well and take it as the standard for expected
behavior. In addition to setting an example by your own approach to
work, you can and should be clear with the staff about how you expect
them to comport themselves as they do their work.

The importance of establishing a good work ethic and valuing scientific rigor should be obvious. However, the importance of how the
organization relates to the public and to other agencies may be less obvious. Under difficult circumstances, many people are reluctant to be
completely open and honest with the public. For example, when we
discovered a few years ago that one of our hatcheries was contaminated
with polychlorinated biphenyls (PCBs; see *A Day in the Life of a Fisheries Administrator*), we feared telling the public all we knew about the
situation. We overcame that reluctance and were very forthcoming with
the public. As a result, we actually gained credibility and public confidence in our decisions regarding how the clean up would proceed. If we
had tried to cover up or gloss over some of that information and it had
subsequently become public, we would have sowed the seeds of distrust
and probably would have lost public confidence. As the first author

researched the characteristics of effective state fish and wildlife agencies, he found many cases where openness and honesty helped agencies overcome potential public relations disasters (McMullin 1993).

Open and honest communication also enhances relationships with other state and federal agencies. Conflicting missions and overlapping "turf" frequently result in some distrust of other agencies. The administrator often sets the tone for how the organization deals with other agencies. Although many approaches may be possible and productive, as an administrator, you will play an important role in determining the approach your organization will take in these relationships.

Keeper of the Program's History and Culture

The fisheries programs in most agencies have long histories, often longer than 100 years. Over that time, many incredible successes, failures, and characters have defined the program. New biologists should be aware of the successes and failures of the past so they understand why the program is where it is today, the hard lessons that have been learned, how they were learned, and where the program should be headed in the future. They should feel that they are part of a long and good tradition of hard work, dealing honestly with the public, and making decisions that are in the best interest of the fisheries resources and the stakeholders of the agency.

The administrator has a unique responsibility to keep that history alive and to ensure that employees are aware of it. Strategies for keeping the organization's history alive include obtaining oral histories from retired employees, inviting retirees to fisheries program meetings to share their experiences and knowledge, and preparing a history of the division. It is only recently that the second author realized the importance of this role as keeper and disseminator of the program's history and culture. If the importance of this role had not been pointed out to him by a former division administrator, he probably would have overlooked it. We believe it is one of the most important things for an administrator to do for the overall health and future of the program.

A Day in the Life of a Fisheries Administrator

Recently, the second author received a request to speak to an upper level class of fisheries and wildlife students on the topic of being a fisheries division administrator. He assumed that the students had little idea of what an administrator actually did on a day-to-day basis, so he kept track of all the things he did one day and related it to them. It was a typical day in the life of a fisheries administrator.

The day began with a meeting with the hatchery bureau chief, engineering staff, and budgeting and accounting personnel. The issue was how to fund nearly $1,000,000 in unanticipated expenses related to cleaning up a hatchery that was contaminated with PCBs from paints that had been used to paint the raceways in the 1960s. Earlier in the year, nearly 800,000 rainbow trout from this hatchery had to be destroyed because of the PCB levels in their flesh.

Next, he met with the legal staff to discuss a lawsuit by a former employee who had been discharged. The meeting participants discussed the merits of the case, as well as various settlement options, including weighing the pros and cons of continuing the lawsuit versus reaching a settlement with the former employee.

The morning ended with a discussion with a regional fisheries manager and biologists regarding the merits of a suggestion to stock forage fish in a reservoir. The issue was complicated by the fact that the predator population was illegally introduced a decade ago and had consumed the existing prey base, including most of the trout the agency stocked annually. The discussion ranged from whether the agency should stock a nonnative forage fish, to which species of forage fish would provide the greatest benefit, to the possible repercussions of doing nothing. The consensus opinion of participants in this meeting was that if the agency did not stock forage fish, the people who stocked the predator fish would likely stock a forage species—and probably not the one the agency would have selected.

Most of the afternoon was consumed by a lengthy discussion with the U.S. Fish and Wildlife Service over development of a Candidate

Conservation Agreement with Assurances (CCAA) for fluvial Arctic grayling *Thymallus arcticus* in the Big Hole River. A petition to list fluvial Arctic grayling as threatened under the ESA had been pending and a lengthy drought was putting additional stress on the population. The purpose of a CCAA is to encourage landowners to take action to benefit an imperiled species before it is listed, in exchange for which they receive some regulatory relief under the ESA if the species is listed. This agreement between the State of Montana, private land-owners, and the U.S. Fish and Wildlife Service was likely to be the fodder for lawsuits and had to be carefully crafted.

At the end of the day, he responded to two questions from the office of the governor. The governor's office had received a complaint from a citizen about evening electrofishing conducted by our crews. This involved explaining the concept of population monitoring to the governor's staff, as well as a discussion of the safety of putting electricity in the water and what it does to fish. The second question originated with a legislator and concerned the agency's community fishing pond grant program.

Attributes of Good Leaders and Administrators

As we mentioned earlier in this chapter, administration often is associated with leadership. A young fisheries professional interested in developing his or her leadership skills could read a different book about leadership every week throughout his or her career and still barely scratch the surface of the leadership literature. Leadership has been a fruitful topic of research, as well as one of the most frequent topics of popular literature. Selznick (1957) suggested that good leaders make the transition from administrative management to institutional leadership (i.e., creating a unique sense of purpose and belonging). Bennis and Nanus (1985) described a transformative leader as one who "... commits people to action, who converts followers into leaders, and who may convert leaders into agents of change." DePree (1992) described leadership as the need to connect one's voice with one's touch (i.e., connecting ideas with actions). Covey (1989) described seven habits of highly effective people: 1) be proactive; 2) begin with the end in mind (i.e., set goals); 3) put first things first (i.e., prioritize your activities); 4) think win–win; 5) seek first to understand, then to be understood; 6) synergize (i.e., look for opportunities to cooperate); and (7) sharpen the saw (i.e., strive to be a life-long learner). These are but a few of the thousands of authors who have contributed to the body of literature on leadership. McMullin and Wolff (1997) described 15 attributes of effective leaders in fish and wildlife agencies.

Our purpose in this section of the chapter is not to summarize the voluminous literature on leadership and administration, but instead, to reinforce many of the points made earlier in this chapter by focusing on the attributes of effective leaders and administrators in natural resource management organizations. We summarize common themes from the literature and results of many informal interactions with natural resource professionals who have participated in leadership development workshops in ten attributes of effective leaders and administrators. If you aspire to positions of leadership, we suggest that you develop these attributes to the best of your ability.

1. Learn to become an effective communicator—Effective leaders and administrators nearly always are able to express themselves clearly both orally and in writing. They can be good salespersons when it is necessary, but the most distinguishing characteristic of their communication style is that they listen as well as they speak. When they communicate with stakeholders, they have a knack for explaining technical information in plain, easy to understand English. Given the rapid pace of diversification in American society, future administrators may need also to explain technical information in Spanish or other languages. As an administrator, you will be called upon to speak often and to many different groups. Whether you are talking to anglers, students, employees, or legislators, your success in communicating usually will depend on how clearly and simply you convey the message. Effective communicators learn how to deliver messages. They know which messages must be delivered in a face-to-face discussion and which messages can be delivered via the chain of command, through written communication, or via email. The ability to speak frankly and to listen well helps to build trust between the leader or administrator and others.

2.   <u>Learn to work with others</u>—Although individual accomplish-ments will help to advance your career, organizations increasingly judge people by their ability to work with others in team settings. An effective administrator leads many teams but also has gained experi-ence in how to be a good team player as well as a leader. Part of being a good team member includes acquisition of good facilitation and meeting management skills. Most definitions of leadership in-clude language about getting things done through others and mobi-lizing people to achieve a common goal (Kouzes and Posner 2003). In general, the further you progress up the organizational ladder, the more you will rely upon others to accomplish your goals. Making the transition from doing it yourself to getting it done through others is one of the most difficult personal tasks for young professionals. Effective leaders and administrators develop their interpersonal skills and learn to read other people (e.g., psychological type). This helps them to communicate more effectively and to "push the right moti-vational buttons." They usually have a genuine interest in working with others.

3.   <u>Be willing to lead by example</u>—Leading by example includes many facets. For example, effective leaders frequently assume a greater sense of responsibility for the success of the organization than other members. They tend to have a sense of the greater good and often will take on tasks and make personal sacrifices to con-tribute to the organization (e.g., financial sacrifices such as relo-cating and taking on administrative responsibilities despite inad-equate financial rewards for doing so). Our experience suggests that effective leaders take on these challenges and make these sac-rifices, at least in part, because they maintain an optimistic outlook about life and their jobs. They believe that they can make a differ-ence and they are willing to try. We have found that cynics rarely change things for the better. Leading by example also means be-ing willing to do anything you ask others to do. Effective leaders know their strengths and their shortcomings. They can be critical in a diplomatic way and also handle criticism of their own work well. As we mentioned previously, effective leaders often lead by example through mentoring of subordinates. A lesser known as-pect of leading by example is taking care of yourself by maintain-ing a healthy lifestyle and effectively managing the stress that in-

evitably goes with administrative positions—and encouraging co-
workers to do likewise.

4.   Focus on the future—To succeed in administration, you must
have a big-picture view of life. Strategic thinking, having a vision
of where you think a program should go, and an idea of how to get
it there, are critical skills for an administrator. With such an outlook, it is not surprising that administrators often tend to be planners. While some administrators can be effective by focusing in-
tently on their work units, the best of administrators tend to think
about what is best for the organization as a whole, while also protecting the interests of their work units.

5.   Embrace adaptability and innovation—Although change is inevitable and necessary for organizations to thrive, most organizations and many of their members resist it. Effective leaders tend to
be more comfortable with change than many other members of the
organization and they may experience frustration with coworkers
who adopt an attitude of, "Why should we change, we have always done it this way?" For example, when presented with a good
idea that is outside of the usual way of doing business, a good
administrator usually looks for a way to put the idea into practice,
while a poor administrator (or the change-resistant coworker) adopts
a position that, "The rules say we can't do that." Perhaps this aspect of leaders' personalities is related to their generally optimistic
outlook on life—they see change as possibly leading to something
good, while others view it as leading to something bad. Leaders
often view change as an exciting challenge and they must exercise
caution not to get too far out in front of their troops. Don't try to
change the whole organization at once. If you want to change
how the organization does something, start at whatever level you
can personally affect. As a regional fisheries manager, the first
author had a different philosophy than many of his peers with regard to planning and public involvement in fisheries management.
Rather than take on the whole establishment, he experimented with
a new approach in his region. His success was noticed and eventually led to a change in the entire fisheries program. Changing
the established way of doing things involves taking some risks and
making some mistakes along the way. The challenge for leaders is

to encourage subordinates to be creative without fear of being punished when mistakes happen. At the same time, responsible risk taking requires that leaders clearly define the bounds of acceptable risk.

6.   Understand the culture and politics of your organization—Many fisheries professionals avoid administration because they do not want to deal with politics. The most cynical of fisheries professionals are convinced that all politics are evil and any decision made by higher-ups in the organization that does not conform to their opinions is politically-based and compromises the resource. However, fisheries conservation is an inherently political process. If you can't accept that politics is a normal part of the process, then administration probably is not for you. Success in the political realm depends on many factors, but among the most important are understanding the history and culture of your organization and maintaining your integrity and credibility in all your dealings with others. Understanding the history and culture of your organization provides the context for how it has operated in the past and the constraints within which it must operate in the future. Integrity is impossible to repair and exercising honesty can require courage. Effective administrators are able to tactfully "tell it like it is." One fisheries administrator we know compares dealing with legislators to a doctor's relationship with his patients. It is difficult for a doctor to tell a patient when he has been diagnosed with a serious illness, but it would be unethical to avoid doing so. Don't make the mistake of trying to tell politicians what you think they want to hear. If they perceive that you lack the courage to tell the truth, you will lose their respect.

7.   Learn how to plan, budget, and get organized—Every organization has its unique system for planning and budgeting. Surprisingly, many employees understand planning, and especially budgeting systems, poorly. Taking the time to learn how those systems work will benefit you as resources are allocated and being in the position to make those allocation decisions allows you to have great impact on conservation of fishery resources. One way to learn is to volunteer to serve on   planning teams and, if the opportunity arises, to assist your supervisor with budget develop-

ment. Although we all know of poorly organized people who are in leadership positions, in general, the further up an organizational ladder you climb, the more important being organized becomes. Perhaps the most important aspect of getting organized is learning how to effectively manage your time to get important tasks done when many urgent, but less important, tasks are demanding your attention.

8.   Learn how to balance decisive action with delegation of authority—It is easy to confuse decisiveness with autocracy; however, effective leaders learn how to appropriately involve subordinates in decision making and to delegate authority. Having the authority to make important decisions is one of the perks of being an administrator, but effective administrators quickly learn that implementation of important decisions also generally requires the commitment of employees. Employees are more likely to be committed to decisions that they had a role in making. Administrators also must be accountable for their decisions. Some employees avoid taking responsibility (and therefore accountability) for making decisions. Those employees are not likely to become effective administrators.

9.   Learn how to evaluate and hire good people—The most important legacy of any administrator is who that administrator hires and promotes. We have participated in numerous hiring processes throughout our careers and none of them involved easy decisions about whom to hire. When you evaluate potential members of your staff, you must balance the need to hire someone with whom you feel comfortable with the need to hire someone who contrasts and complements your strength and weaknesses. Administrators walk a fine line between hiring people who will be good team members and hiring people who think exactly as they do.

10.   Remember why you got into the profession—Nearly all of us have, at one or more times in our careers, stopped during a work day and thought, "I can't believe I am being paid to do this!" As your career progresses, you may have fewer of those moments, but it is important to recreate them occasionally. As your career

progresses, take advantage of the occasional opportunity to do some of the enjoyable tasks you did more frequently early in your career. This is one way to stay in touch with the world your staff deals with and, at the same time, rekindle the passion for conservation that all true fisheries professionals possess.

Some Concluding Thoughts

Effective administrators are just as important to fisheries conservation as the scientists who collect the information upon which conservation decisions are made. It is unfortunate that so many fisheries professionals shy away from careers in administration because the profession deserves to have its best people move into positions of leadership.

While we encourage young fisheries professionals to consider administration as a career goal, we also admonish them not to be in a hurry to climb the organizational ladder. Each position has its unique opportunities to learn and enjoy the profession. Every time you consider applying for a new position, you should ask yourself if you still are acquiring new knowledge and skills in your present position and if you are enjoying your present position too much to move on. In addition, you should consider whether you are ready to take on a new challenge. You will be a better administrator when the time arrives if you have gained a variety of experiences along the way. One of the criteria we have seen used in making a hiring decision is whether the candidate has 20 years of experience or 1 year of experience, 20 times over. There should be no shortage of opportunities to move up during the first two decades of the 21st century, as 75% or more of current administrators in federal and state agencies are members of the Baby Boomer generation who will be retiring during that period (Colker and Day 2003; McMullin 2004). If you develop the skills and attributes of effective leaders and administrators discussed previously in this chapter and gain a variety of experiences in your career, you will be well prepared to move into administration.

Finally, we also advise young fisheries professionals, including those who cannot foresee ever wanting to be an administrator, never to say never. Neither of the authors of this chapter entered into the profession with the intention of becoming administrators. The first author moved from state agency administration to academia in mid-career, and that after once declaring that he would never seek a Ph.D. because he was sure that his career goals would not require the additional education!

References

Bennis, W., and B. Nanus. 1985. Leaders: the strategies for taking charge. Harper & Row Publishers, New York.

Colker, R. M., and R. D. Day, editors. 2003. Conference on personnel trends, education policy, and evolving roles of federal and state natural resource agencies. Renewable Resources Journal, Winter 2003–04.

Covey, S. R. 1989. The 7 habits of highly effective people. Free Press, New York.

De Pree, M. 1992. Leadership jazz. Doubleday, New York.

Kouzes, J. M., and B. Z. Posner. 2003. The leadership challenge, 3rd edition. Jossey-Bass, New York.

McMullin, S. L. 1993. Characteristics and strategies of effective state fish and wildlife agencies. Transactions of the North American Wildlife and Natural Resources Conference 58: 206–210.

McMullin, S. L. 2004. Demographics of retirement and professional development needs of state fisheries and wildlife agency employees. Report prepared for the U.S. Fish and Wildlife Service National Conservation Training Center. Virginia Tech, Blacksburg.

McMullin, S. L. 2005. Baby boomers and leadership in state fish and wildlife agencies: a changing of the guard approaches. Transactions of the North American Wildlife and Natural Resources Conference 70: 27–37.

McMullin, S. L., and S. W. Wolff. 1997. Preparing tomorrow's fish and wildlife agency leaders. Fisheries 22(2): 24–25.

Selznick, P. 1957. Leadership in administration: a sociological interpretation. University of California Press, Berkeley.

Chapter 14

Equal Opportunities in Fisheries Employment

COLUMBUS H. BROWN AND
ESSIE C. DUFFIE

Workforce diversity is extremely important in today's society and demographic projections indicate a rapidly increasing minority population. According to the 2000 U.S. Census, approximately 30% of the nation's population currently belongs to a racial or ethnic minority group. Projections show that minorities will constitute almost half of the U.S. population by mid-century. Much of this population growth is occurring in areas where better fisheries management is needed to restore populations of imperiled fish and foster sustainable stocks of harvested species. However, in order to make effective management decisions about resources in these areas, scientists and managers should reflect the diverse populations they serve (Cuker 2001).

Demographic trends raise concerns about the ability of fisheries and other scientific professions to meet the needs of their future workforce. Historically, non-Hispanic white males have dominated these professions with very poor representation by minorities and women. The representation of minorities, specifically African Americans, Hispanics, Native Americans, and women among science educators and students at most academic institutions still lags behind their representation in society at large (NSF 2004; Handelsman et al. 2005). Asian Americans are a small percentage of the population in the United States; however, they are not generally considered to be underrepresented in the sciences. Although progress has been made and diversity does exist in some academic institutions and private and governmental natural resources organizations, there is still insufficient minority representation overall in fisheries disciplines. As fisheries science and management have evolved to reflect the complex and interdisciplinary nature of resource issues, there is a clear need to involve people with a variety of backgrounds, cultures, experiences, and skills. It is essential that future cohorts of fisheries professionals include individuals from underrepresented groups that are educated in issues of natural resources management. Indeed, if a greater proportion of fisheries professionals are not recruited from the growing population of minorities and underrepresented groups, there may be a serious shortfall in the number of adequately trained fisheries professionals.

Furthermore, it is unlikely that fisheries professionals will be able to adequately address the needs of a more diverse public without a more diverse workforce. For example, one of the most evident needs is the ability to communicate effectively with the growing number of Americans for whom English is not their primary language. The necessity for agencies to publish regulations and information in Spanish and other languages continues to grow (Henderson 2004). In addition, assessing the needs of a diverse public will require a more intimate understanding of their attitudes and behaviors toward fisheries resources (e.g., consumption vs. conservation). Human dimensions research has begun to refute some of the stereotypes about minorities and women that fed into management decisions, such as the reasons that these groups do not participate as actively as other users of fisheries resources (Duda et al. 1998). Detailed information about the patterns of participation and expenditure of minority and women hunters and anglers has revealed that these groups are unique in many respects (Henderson 2004).

Resources and Opportunities

Numerous opportunities are available for individuals from underrepresented groups who are interested in careers in fisheries and related professions. In fact, people taking advantage of existing resources and opportunities has been one of the major reasons for increases in the number of minorities and women hired for fisheries-related jobs. In this section, we describe a variety of available resources and opportunities, including internships, scholarships, fellowships, volunteer opportunities, and participation in scientific conferences and professional societies. The most successful individuals will take advantage of a number of these in combination to enhance their competitiveness in the job market.

Internships

Internships give students a chance to learn firsthand what a career in fisheries involves. In most instances, students will be paired with a mentor who guides them throughout the experience. Students are usually provided with a stipend to cover basic living expenses and often receive housing assistance as well. They gain valuable experience in the field and laboratory and may participate in any aspect of fisheries science, from collecting data to writing reports. Internships not only provide an opportunity for students to add credible work experience to their resumes, but it also helps them begin to chart a course for a successful career. The American Fisheries Society (AFS) provides access to a substantial number of internships and other short-term job opportunities for students through its affiliations with public and private organizations. In particular, AFS administers the Hutton Junior Fisheries Biology Program (see Chapter 1), which was designed to stimulate interest in fisheries careers among high school students from groups that are underrepresented in the fisheries profession, including minorities and women.

Scholarships

Scholarships offer financial assistance in the form of grants or tuition support to students on the basis of achievement, ability, or financial need. Scholarships are generally awarded to those students who have demonstrated their abilities, have received certain types of training, or exhibit qualifications that would make them an asset to an employing agency or organization. Several government sources provide scholarship funds each year both for general education purposes and for students in fisheries-related fields of study. The American Fisheries Society in particular has made efforts toward increasing diversity in the profession and the Society by conferring special awards. For example, the

Equal Opportunities Section of AFS administers the J. Frances Allen Scholarship Award, which is given to a female doctoral fisheries student and was established to encourage women to become fisheries professionals.

Fellowships

Fellowships are a form of financial assistance given to students based on academic achievement and are intended to support students as they continue their studies. Fellowships are most commonly given to graduate students. A variety of fellowships are available through organizations and agencies in the U.S., including the National Science Foundation, the National Academy of Sciences, the Environmental Protection Agency, individual academic institutions, and private organizations. Fellowships are typically granted for one to a few years and vary from a few hundred to many thousands of dollars.

Volunteer Opportunities

Volunteering to work on a project being conducted by fisheries researchers is a great way to gain valuable work experience. Such experience is always important as part of your list of experiences on your resume. Volunteer experiences are similar to internships except that volunteers are not paid, and thus they demonstrate professionalism and a sincere drive to pursue career goals. Most any fisheries agency or organization will make volunteer opportunities available to interested individuals.

Scientific Conferences

Attending scientific conferences is critical to becoming a solid fisheries professional. Each year, the AFS Annual Meeting provides education, training, networking, and mentoring opportunities for a diverse audience. The meeting is an excellent forum through which fisheries students and professionals can obtain information on the latest advances in the field and current employment opportunities. The Equal Opportunities Section works to increase participation of minorities and women in the Society and the profession, and provides travel grants to students from underrepresented groups to attend the AFS Annual Meeting. The Section also sponsors a luncheon at the meeting designed to foster mentoring of students from underrepresented groups.

Professional Societies

Active membership in professional societies demonstrates dedication and can provide assistance throughout your career. Chapter 15 offers a full discussion of the ways in which AFS can foster professional development. In addition to AFS, the American Society of Limnology and Oceanography (ASLO) operates a special *Minorities in the Aquatic Sciences* program to recruit and train minorities interested in aquatic sciences, including fisheries. The advantages provided by professional society membership are numerous and may include discounts on publications, increased professional credibility, and training and networking opportunities. It is crucial that individuals from underrepresented groups become active members of professional societies if they are to become fully integrated into the profession.

Student Career Experience Program

The Student Career Experience Program (SCEP) is part of the federal Student Educational Employment Program and was established to:

1. Recruit exceptional employees into federal agencies;

2. Support equal opportunity employment objectives;

3. Provide exposure to public service; and

4. Promote educational opportunities.

This is an outstanding program that has provided many minority and women federal employees with their start in fisheries and related professions. Through SCEP, agencies appoint students to serve in "trainee" positions that provide them with valuable work experience in their field of study. Once they complete their degrees, students who meet the minimum number of required work hours in the program can be noncompetitively converted to career or career-conditional positions within the agency. Student Career Experience Program appointments are available to students from high schools, accredited technical and vocational schools, 2- or 4-year colleges and universities, and graduate and professional schools. To be eligible for the program, students must be U.S. citizens, pursuing fields of study related to the work of the employing agency, must be at least the minimum age (required by federal, state, or local laws and standards) governing the employment of minors, and must be taking a minimum of a half-time course load.

Mentoring and Retention

Effective mentoring of new employees is critical to helping them be productive, turn out quality work, and thrive in the workplace. Similarly, the working environment needs to be welcoming and accepting of the backgrounds and strengths of employees in order for them to be successful. Employees from underrepresented groups provide much-needed diversity in the workplace, and efforts should be made to ensure that they are comfortable, effective members of the team. Furthermore, because new employees require considerable investments of time and money for hiring, orientation, and training, retaining employees is important to sustaining productivity. Employers should seek to provide a healthy work environment so that employees can continue to fully apply their talents in a meaningful way. Importantly, employees from underrepresented groups that have settled in and become prosperous members of an organization will also be especially prepared to mentor new employees from similar groups.

Legal Issues

Equal Employment Opportunity (EEO) laws have affected the way companies and organizations in the United States conduct business, especially when it comes to policies and practices related to recruiting, hiring, training, evaluation, transfers, promotions, and pay. Additionally, affirmative action is a tool designed to support the employment of individuals from groups that have been discriminated against in the past (Crosby and VanDeVeer 2000). Affirmative action occurs whenever a company or organization goes out of its way to make sure that there is no discrimination against minorities, women, people with disabilities, or veterans. It is intended to correct underutilization of qualified women and minorities, but is not intended to create preference for them to the exclusion of other groups.

Rights of Applicants

Equl Employment Opportunity laws prohibit discrimination and guarantee a level playing field for all candidates during recruitment and employment. Any applicant who meets the required qualifications listed in a job advertisement or position description has a right to be examined further based on his or her qualifications. Furthermore, an applicant has a right to be informed by the hiring agency or institution of the process by which a selection will be made.

Agencies and institutions often appoint search committees to fill job vacancies. These committees should comprise a diverse array of individuals that will interact with the successful candidate. The interview process should define for the applicant the essential qualifications required by the job and may involve discussion of other desired qualifications, but care must be taken to ensure that all applicants are treated in an unbiased manner. Applicants should not be asked any inappropriate questions, such as those relating to their personal life or sexual orientation.

Applicants and employees of entities under federal legal jurisdiction are protected under federal authority and have a right to file a complaint of discrimination. No one can be subjected to retaliation for opposing any practice made unlawful by an EEO statute (see *Legal Requirements for Employers*). In particular, gender equity complaints involving disputes over pay and promotion are covered under the Equal Pay Act and prohibitions against sexual discrimination that cover sexual harassment are included in Title VII of the Civil Rights Act. Sexual discrimination includes practices ranging from direct requests for sexual favors to workplace conditions that create a hostile work environment.

Individuals should contact the appropriate authorities if they be-

lieve that EEO laws or procedures have been violated. Most companies have human resource officers and federal and state agencies have
EEO or civil rights personnel who can help guide applicants or employees through this process. The Equal Employment Opportunity
Commission (EEOC), which enforces EEO laws and provides oversight and coordination of all federal EEO regulations, can also be of
assistance. If a number of people have similar complaints, class-ac-
tion procedures can be used. Other options for assistance include seeking legal counsel, which is available on most college and university
campuses, contacting state bar associations through Legal Aid Associations, contacting the Office of Federal Contract Compliance Programs in the Department of Labor (if the complaint is against a federal
contractor), and contacting the American Civil Liberties Union (ACLU).
The ACLU is a private organization that represents the interests of all
employees and citizens and has regional offices located throughout
the United States.

Legal Requirements for Employers

All federal, state, and local government agencies, as well as companies and institutions that receive federal funding (including academic
institutions), are under legal jurisdiction of federal statutes. Equal opportunity employment is mandated by federal law, so it is unlawful for
these agencies and institutions to discriminate against anyone during
the hiring process, the term of employment, or as part of evaluation
procedures. It is the policy of the federal government to prohibit discrimination based on race, color, religion, gender, national origin, age,
disability, or status as a Vietnam veteran.

Major civil rights mandates and statutes that support EEO include,
but are not limited to, the following: Titles VI and VII of the Civil
Rights Act of 1964 (as amended by EEO Act of 1972); Equal Pay Act
of 1963; Title IX of the Educational Amendments of 1972; Rehabilitation Act of 1973 (sections 501, 503, 504, and 505 as amended); the
Vietnam-Era Veterans Readjustment Assistance Act of 1974; Age Discrimination in Employment Act of 1967 (as amended in 1974); Americans with Disabilities Act of 1990; and the Civil Rights Act of 1991. In
addition, Executive Order 11375 prohibits discrimination in employment based on gender by federal contractors or subcontractors who

employ 50 or more persons with $50,000 or more in contractual obligations.

Future Directions

There has been, and continues to be, considerable debate over gender, race, and merit with regard to programs designed to increase participation by minorities and women in science (e.g., Crosby and VanDeVeer 2000; Summers and Hrabowski 2006). Regardless of the outcomes of this debate, equal opportunity employment is necessary to build diversity in fisheries and related professions and should be a fundamental value of any organization. Women and individuals from minority and other traditionally underrepresented groups must be recruited in order to remain competitive and relevant during times when the scope and nature of these professions are changing so rapidly. Employers should keep abreast of legal requirements and reach out to encourage young people from diverse backgrounds to enter the profession. At the same time, women and individuals from minority and other underrepresented groups should take full advantage of available resources and opportunities.

Sources of Additional Information

The American Fisheries Society maintains a list of resources for students, including internships, scholarships, fellowships, and volunteer and employment opportunities, in the *Student* section of its web site (http://www.fisheries.org). Current information about the AFS Annual Meeting, which is held in the late summer or fall of each year, is also available on the AFS web site. The Equal Opportunities Section of AFS has a separate web site with additional information (http://www.fisheries.org/eos). Students interested in the *Minorities in the Aquatic Sciences* program of ASLO should consult the web sites maintained by ASLO (http://aslo.org/mas) and Hampton University (http://www.hamptonu.edu/academics/schools/science/marine/aslo). The Association for Women in Science maintains a web site (http://www.awis.org), which has a wealth of information including instruc-

tions for ordering their report, *A Hand Up*. Information about the federal Student Career Experience Program can be found on the web site
of the Office of Personnel Management (http://www.opm.gov/employ/
students). Information and assistance available through the U.S. Equal
Employment Opportunity Commission (http://eeoc.gov) and the Office of Federal Contract Compliance Programs (http://www.dol.gov/
esa/ofccp) can be accessed through their web sites.

References

Crosby, F. J., and C. VanDeVeer, editors. 2000. Sex, race, and merit: debating
affirmative action in education and employment. University of Michigan
Press, Ann Arbor.

Cuker, B. E. 2001. Steps to increasing minority participation in the aquatic
sciences: catching up with shifting demographics. American Society of Limnology and Oceanography Bulletin 10(2): 17–21.

Duda, M. D., S. J. Bissell, and K. C. Young. 1998. Wildlife and the American mind:
public opinion on and attitudes toward fish and wildlife management. Re-
sponsive Management, Harrisonburg, Virginia.

Handelsman, J., N. Cantor, M. Carnes, D. Denton, E. Fine, B. Grosz, V. Hinshaw, C.
Marrett, S. Rosser, D. Shalala, and J. Sheridan. 2005. More women in science.
Science 309: 1190–1191.

Henderson, E. 2004. Participation and expenditure patterns of African-American,
Hispanic, and female hunters and anglers. Report 2001-4, addendum to the
2001 national survey of fishing, hunting, and wildlife-associated recreation.
U.S. Fish and Wildlife Service, Arlington, Virginia.

NSF (National Science Foundation), Division of Science Resources Statistics.
2004. Women, minorities, and persons with disabilities in science and engineering: 2004. NSF 04-317. NSF, Arlington, Virginia. Available: http://
www.nsf.gov/statistics/wmpd. (April 2006).

Summers, M. F., and F. A. Hrabowski, III. 2006. Preparing minority scientists and
engineers. Science 311: 1870–1871.

Chapter 15

The American Fisheries Society: A Framework for Professionalism

GHASSAN (GUS) N. RASSAM

Membership in a professional society is a good indicator of an individual's commitment to professionalism. Like other professional societies, the American Fisheries Society (AFS) offers its members many opportunities for career and leadership development, for communication among peers, and for advancing common interests. The American Fisheries Society mobilizes the expertise of its members to address scientific and management concerns, pools the voices of its members to speak with strong effect on aquatic resource issues, and generates publications, symposia, workshops, and continuing education courses to help keep the fisheries profession up-to-date and effective. The American Fisheries Society also helps maintain and improve standards of performance and public responsibility within the profession, and encourages high quality work through an extensive awards and recognition program. In turn, AFS works to enhance the public stature of the fisheries profession and to increase public support for fisheries science and management.

To say that AFS "does" all these things is to say that its members do them. The American Fisheries Society is simply an organization that fisheries professionals have created to magnify their individual efforts and to improve themselves and others as scientists and stewards of fisheries resources. Every fisheries professional has a stake in at least some of the issues that AFS addresses. Therefore, every member ought to help define and participate in AFS activities. The organizational diversity of AFS allows it to support member activities at local, regional, national, and international levels, as well as within specific disciplines (Sections). The American Fisheries Society currently offers more opportunities for its members to participate in professional activities than it ever has before.

Similar to other professional societies, the benefits of participating in AFS and the opportunities afforded by that participation accrue at every stage in an individual's career (Box 15.1). As the rising generation of fisheries professionals, students begin to establish their network of professional relationships by attending AFS Annual Meetings or meetings of other AFS units—chapters, divisions, and sections. As their research develops, they can showcase their work and themselves to critical, but supportive, audiences at these meetings and gain valuable experience in technical communication. By participating in AFS units and serving on volunteer committees, students can broaden their exposure to professional issues and take advantage of opportunities to assume responsibility and exercise leadership. Membership in AFS also gives students greater access to scholarships, fellowships, grants, and jobs, further aiding their growth from apprentice to professional. Recent changes to the membership dues structure of AFS offer students membership at a quarter of the regular membership price, and with it free access to online publications. The online publications include AFS journals and *Fisheries* magazine, with electronic versions (PDF) of articles from all of the publications from their inception to the present (older material has been archived in electronic format through the Fisheries InfoBase project). In addition, young professionals pay only half of the regular membership dues for a period of 3 years following graduation. Professionals from developing countries, a category that is broadly defined, pay only nominal entry fees to become members of AFS.

Entry-level to mid-career fisheries professionals find AFS membership an increasingly valuable asset. The American Fisheries Society meetings provide opportunities for them to steadily expand their networks of colleagues and information sources. In addition, continuing education courses and special workshops that are offered at most AFS meetings allow them to keep up with the latest developments in science and management. A recent sampling of continuing education courses covered topics from Geographic Information Systems and te-

Box 15.1. Opportunities and benefits provided by the American Fisheries Society that professionals can take advantage of at various stages of their career.

Career Stage	AFS Opportunities and Benefits
High School	*Careers in Fisheries* brochure Hutton Junior Fisheries Biology Program Mentoring through a local AFS Chapter Low membership dues
College	*Careers in Fisheries* brochure Mentoring through a local AFS Chapter Low membership dues Student Subunits *Fisheries* magazine Free access to online publications, including journals
Graduate School	Low membership dues Student Subunits *Fisheries* magazine Free access to online publications, including journals Student Subsection of the Education Section Section membership Skinner Travel Award Other AFS travel awards and scholarships Low registration fees for meetings Networking opportunities at meetings Leadership training and continuing education courses at meetings Reasonably priced textbooks Journal publishing
Young Professional	Low membership dues *Fisheries* magazine Section membership Networking opportunities at meetings Leadership training and continuing education courses at meetings Low prices on books Journal publishing AFS Professional Certification

Box 15.1. Continued.

Career Stage	AFS Opportunities and Benefits
Professional	*Fisheries* magazine Section membership Networking opportunities at meetings Leadership training and continuing education courses at meetings Low prices on books AFS Professional Certification World Fisheries Congress
Retired Professional	Low membership dues *Fisheries* magazine Retired members travel award to attend the AFS Annual Meeting Low registration fees for meetings Networking opportunities at meetings Friendships

lemetry to leadership development and scientific writing. The American Fisheries Society journals and books on a similarly wide variety of topics allow professionals avenues through which to disseminate their scientific work. Publication outlets, symposia, workshops, and continuing education courses help them maintain and hone their technical edge and, as they move into supervisory positions, their managerial skills. If members cannot find what they need professionally, AFS often can help them create it. Individuals at almost any stage in their career can have their qualifications peer-reviewed through the AFS Professional Certification Program. In fact, more and more fisheries professionals from all over the world are finding that certification by AFS enhances their status and compensation in the workplace. Professionals who work for a public agency in the United States or Canada will also value the periodic AFS surveys of fisheries salaries in state, provincial, and federal governments. Indeed, past surveys have led to improved compensation in some instances. Whether their interests lie in technical specialties or broad resource policies, established professionals can find other AFS members, and usually an AFS unit or committee, to help them further those interests. Perhaps most importantly, their sustained participation in AFS earns them growing responsibility and stature within AFS, and thus within the profession. This allows them to exert greater influence on the science-based management of aquatic resources. Furthermore, as they acquire leadership skills their judgments and counsel are increasingly valued in the marketplace of ideas.

Experience and professional recognition are cumulative, and senior professionals typically have the best opportunities to affect fisheries policy at national and international levels. If they have been active in AFS, and especially if they have held leadership positions in AFS units or served on committees, such professionals will better understand how to mobilize AFS resources to accomplish professional objectives. By doing so, they can leverage their personal capabilities in a substantial way. Even if they do not choose leadership roles, senior professionals can influence AFS policies and programs through the wisdom of their counsel and the mentoring they offer younger members. For themselves, continued interactions with both old and new colleagues at AFS functions keep them abreast of the latest fisheries issues. Well after their official retirement, many people remain active and productive in the fisheries profession, in part because of the stimulation and learning they obtained from continued participation in AFS. At any level of professional development, members find that what they get out of their involvement in a professional society is directly proportional to what they put into it. Sustained involvement in AFS will bring the greatest rewards.

AFS Professional Certification

Certification in the AFS Professional Certification Program is based on peer evaluation of an individual's academic achievement, publications, and relevant professional experience. It is a valuable mechanism for advancing the profession and recognizing achievements in fisheries science. It establishes a benchmark of qualifications and accomplishments that is recognized among AFS members, by courts of law, and increasingly by employers seeking highly qualified professional candidates.

Ecosystems and populations are very sensitive to manipulation by humans, including natural resource managers, and the need to measure and ensure professional competence is particularly acute for resource stewardship. Standardized testing of applicants to determine their qualifications for certification is a need recognized by many, as is a linkage of certification renewal to continuing education and a commitment to life-long learning. All renewable resource professions face the challenge of having certification recognized as a requirement for employment in resource management (The Wildlife Society and The Ecological Society of America also have certification programs). One day these related professions may collaborate to develop one or more standardized certification programs that can be administered by an independent body and be internationally recognized.

A workable certification program must set strong standards, yet be flexible enough to accommodate varying degrees of experience and specialization within a diverse profession. Differences of opinion about certification, such as how and to whom it should apply, and to what standards, make it very difficult to change a broad-based program, as the AFS experience has shown. Efforts to improve certification will be worthwhile if the results of those efforts are that aquatic resources become better managed by scientific and humanistic principles and that ethical standards for resource stewardship become codified in management practices in an enforceable way.

Professional Communication

Professional benefits from AFS or any other source do not arise in a vacuum. Rather, they stem from human interactions and communication. Effective communication is important for career development, but it is also essential for mobilizing others to serve a common cause. Furthermore, fisheries professionals have a public responsibility to communicate their knowledge because the public has invested in their training, will continue to invest in their work, and has entrusted some of its most valued resources to their care. The level of communication should be appropriate to the audience being addressed. Professionals that deal with both science and the application of science to resource management are likely to encounter a diverse set of audiences for professional communications. While the basic message may be the same and must always be based on the best scientific information available, they will need to adapt to the needs of these different audiences in order to be effective. A group of local high school students will require a vastly different approach than a committee of congressmen.

Both informal and formal communications are important professionally. The informal networking that goes on at meetings and by email, phone, and fax is an efficient way to give and receive current information and it cements the personal relationships needed for cooperative activities. As a professional's network grows, so does that individual's ability to cope with developments and influence them. Professionals simply cannot function effectively without networks. The only practical limit on network size is the time available to maintain it.

Formal communications have structured content and logical development. They result from rigorous thinking about a given subject. Their purposes range from channeling action in a certain direction (e.g., fishery regulations) to persuasion (e.g., grant applications), and they include academic lectures, public speeches, essays, white papers, policy documents, legislative testimony, letters and reports, scientific papers, listserv communications, and blogs, among others. They may contain some "raw" information but, for the most part, the information has been summarized, edited, and interpreted with the intent of influencing the behavior or thinking of some target audience. The audience is more or less anonymous (e.g., journal readers, fishing club members, listserv subscribers), different in mind-set from the communicator (e.g., students, senior agency directors), or both.

Formal communications are among the most difficult tasks facing fisheries professionals because of the intellectual rigor they require and the difficulty of foreseeing audience reaction to them. They are essential tasks nevertheless. Taking them on and doing them well will bring professional advancement. Shirking them or doing them poorly can stunt a career. Here are seven tips for effective formal communications.

1. <u>Know the principles of language</u>—They undergird everything,

spoken or written. Bad grammar, tortured syntax, and incomprehensible word strings can destroy a message and debilitate the messenger. Take courses, if necessary, and practice.

2. <u>Tell a story that will be interesting to your audience, but do not "show off"</u>—Remind the audience why they should be interested in your message, but minimize hype. Preview your conclusions early on so the audience can better understand the force of your argument as it develops. Make sure the audience does not get lost as the story unfolds. Introduce information only when it is needed, not before or after. Excise material that does not contribute to the story.

3. <u>Learn as much as possible about the audience in advance</u>—Read the journal to which you intend to submit a manuscript and talk to people who have dealt with the editor. Find out before you speak to a sporting club whether its members fish for walleye *Sander vitreus* or bass, or whether they favor harvest over catch and release. <u>Be sensi</u>tive to the culture of your audience, as it makes a difference whether you are addressing a local audience or the World Fisheries Congress. Determine what the audience already knows and tailor your presentation accordingly. Analyze your experience with each audience so you will be better prepared for the same or a similar audience in the future.

4. <u>Learn the idiom and culture of the audience</u>—Get a feel for the terms and phrases people use, the words that attract or annoy them, and the ideas they accept or question. Use metric units of measure if the culture demands it. Avoid erudite jargon if the audience will not comprehend it. If you challenge a community truth, make appropriate bows to the existing paradigm and make the change seem evolutionary rather than revolutionary. Do not gloat.

5. <u>Respect the audience</u>—Do not communicate over the audience's head or insult their intelligence. Be honest with them and tell them something useful. Do not waste their time.

6. <u>Accept criticism professionally</u>—Be open to differing opin-

ions, whether they come from the audience or from peer reviewers. The majority of criticisms are provoked by bad scripts, not by inherently bad messages. Script problems can be corrected. Try to anticipate problems by consulting members of the future audience. Reviews by friends, spouses, and close colleagues can help a great deal, but realize that such people are members of your culture and not necessarily the culture of your audience.

7. Follow instructions—If a journal editor requires four copies of a manuscript, send four. If a session chair indicates that PowerPoint (Microsoft Corporation, Redmond, Washington) is the only acceptable presentation medium, bring a PowerPoint presentation. Meet posted deadlines. Rules are set to expedite the process of presenting and evaluating information. The people who manage that process often control what information will be presented, so it is risky to annoy them with unnecessary problems.

In the fisheries profession, those who express themselves clearly in the language of their audience earn respect and influence, and their careers advance accordingly. Communication skills should be learned early and honed often. An individual's first manuscript submission, public speech, or job interview will show the fundamental importance of these skills. AFS provides many opportunities for enhancing communication skills and for disseminating information to the right audience. In addition to material posted on the AFS web site and the continuing education courses on technical writing that are provided at AFS meetings, the Fisheries Conservation Foundation maintains a library of PowerPoint presentations that are excellent vehicles for communicating with the public. Information distributed by email over the AFS listserv reaches a great many people in the profession, both members and nonmembers. As a platform for active communication and advocacy for conservation and science, AFS has provided many channels for communicating science in recent years. To pick a recent example, the work of Ted Ames, the long-time Maine fisherman turned scientist and one of the 2005 MacArthur Fellows (also known as the "Genius" awards), came to the attention of the world primarily through a research article he published in *Fisheries* magazine (Ames 2004).

Ethics

Personal honesty and respect for others is at the core of all ethical systems. Standards of personal ethics, which are learned from birth and reinforced in various ways by society, give individuals a clear sense of right and wrong and of how to interact with other people. Respect can be given to institutions and ideas as well as individuals. In a system of professional ethics, the profession itself is the focus of respect. For example, the AFS Standard of Professional Conduct states that members will "strive to preserve and enhance the fisheries profession." In part, they accomplish this through personal honesty in reporting their work, in dealing with colleagues, employers, clients, and the public, in representing their qualifications and expertise, and in resisting coercion of their professional judgments. They also do so by according respect to other professionals, by giving credit for their work, exchanging information, restricting technical criticisms to technical forums, and maintaining confidentiality. Finally, they work to enhance the profession through self-policing, encouraging good conduct in others and exposing misconduct whenever they encounter it.

To some extent, professional ethics are like individual ethics in that they are solutions to problems of security and survival, but they can also embrace altruistic ideals. The AFS Standard states, "First and foremost, on joining the AFS, a member accepts the responsibility to serve and manage aquatic resources for the benefit of those resources and of the public." This level of respect for human and nonhuman systems has not reached the sophistication of Aldo Leopold's land ethic, but it is closer to that ethic than it used to be. Only recently have law and practice begun to codify the principles that humans are an integrated part of ecosystems and that we have a responsibility, in Leopold's words, to "preserve the integrity, stability, and beauty of the biotic community" (Leopold 1949). This ethic serves, and is served by, modern goals of ecosystem management and resource stewardship. Fisheries professionals, who are asked increasingly often to help refit patterns of human activity to patterns of the landscape, will do well to foster this ethic within society at large.

Professional ethics are noble on paper and encourage nobility in behavior and thought. However, few professions have institutions and mechanisms for judging and punishing ethical misbehavior. The fisheries profession does not. Most fisheries people encounter ethical conflicts during their careers, such as unearned co-authorships, falsified data, dishonest testimony, mistreatment of others, and abuses of natural resources. Problems that seem straightforward in some situations are ambiguous in others. An individual's own conscience and the support of respected colleagues may be the only guides to appropriate action in such situations.

As many governmental natural resource decisions are challenged by a variety of stakeholders and by economic pressures, the traditional definition of peer review is sometimes broadened to include non-science stakeholders and the motivations of reviewers are increasingly questioned. In such an atmosphere, it is essential that all professionals adhere to the ethical standards set collectively by professional organizations. In recent years, several natural resource professionals working for federal and state agencies have been accused of "bad science" or "bad ethics." Professional society codes of ethics give the public some assurance that members of organizations, such as AFS, observe certain standards in the conduct of their research and work.

Keys to Success

Larry Olmsted, a recently retired fishery scientist, spoke to students at the Fourth Annual Southern Division AFS Student Colloquium in 2004 about how to get really smart, get a super job, and be wildly successful. His remarks were adapted for publication in *Fisheries* magazine (Olmsted 2005a, b). Based on an informal survey of successful professionals, he noted that respondents had a somewhat difficult time defining success, but offered the following working definition: *"Success is achieving goals that have a lasting impact on society or your profession and align with your personal values. All activities are performed in such a manner that others (peers) recognize not only the results, but view you as a role model for how those results are achieved."* By condensing the survey responses, he provided the following top 10 keys to achieving success, in increasing order of importance:

10.   Be knowledgeable—It seems apparent that it will be difficult to be successful if you are ignorant of important facts or concepts in your area.

9.   Be flexible—Restricting yourself geographically or by job function is a sure way to limit your success, and may even limit employability.   The field of fisheries changes continually and we must be flexible in adapting and succeeding.   A part of flexibility is being willing to admit mistakes and accept advice and criticism. We must dare to make mistakes and not be ashamed to acknowledge them. Criticism can be a valuable tool for self improvement if it is offered and accepted in a constructive fashion.   Seek out those who will provide honest, constructive criticism.

8.   Copy your style from giants—Tom Kwak, an active AFS member and an author of Chapter 7, made the following comments: "There are a few exceptionally talented, experienced, and classy individuals in our field, and if you're fortunate enough to cross their paths, you're crazy not to adapt aspects of their style that work for you.   It's easy to integrate techniques of such giants into your own mode of operations, and yet to maintain your unique personality."   Jim Martin, a long-time fisheries professional and AFS member, stresses the importance of such individuals in our careers and encourages us to thank them.   Who are the giants in your career?   Have you thanked them?

7.   Have fun—If you are not having fun, change jobs. Fun fosters enthusiasm, which is critical for success.

6.   Work hard, persist—We will all encounter obstacles and stumbling blocks in our career.   It is critical how we react to these setbacks.   Many really successful people made plenty of mistakes, and many had failures.   Charlie Liston responded about persistence, reminding us that Dr. Seuss's first children's book was rejected by 23 publishers, that Michael Jordan was cut from his high school basketball team, that Henry Ford went broke five times before success, and that Winston Churchill took three years to get through the eighth grade (he couldn't learn the English language).

Kim Erickson summarized the need for persistence by saying, "Attitude is everything. Life is 10% what happens to me and 90% how I react to it."

5. Be a team player—The basic elements of being a team player are to accept responsibility, share the credit, and communicate effectively. Increasingly, fisheries professionals are operating as part of interdisciplinary teams. It is critical that you understand your role and operate as an effective member of a number of different teams. The days of achieving success while operating as a "Lone Ranger" are gone forever. Almost any project you become involved in will require that you interact and cooperate with other biologists, administrators, legislators, regulators, and a variety of stakeholders. Successful fisheries professionals manage all of these relationships effectively. In the end, your success will be determined more by how you manage relationships with people than by how you manage fish.

4. Build a network—Quoting Mike Van Den Avyle: "Continu-ously identify others in your profession who share similar goals and are fun to work with, and always invest personal energy and time to build, maintain, and expand working relationships relevant to your goals. Do this over as broad a geographical area as possible, and always push beyond your comfort zone to cultivate personal relationships with those in other disciplines, institutions, and working cultures. The breadth and depth of your 'professional friends' is a source of energy, motivation, and knowledge that breeds and accelerates success throughout the development of your career."

3. Maintain balance—If we achieve professional success at the expense of our spiritual, physical, or social goals, it is a victory that comes with too high a cost. Gary Breece noted, "You can't have a successful professional career when your personal life is in shambles, no matter how many streams you save, students you teach, or papers you publish." The spiritual, emotional, and social aspects of your life require continual attention and nurturing, but it is easy to tilt the balance in pursuit of professional goals. Don't allow it to happen.

2. <u>Never compromise your integrity and honesty</u>—Scott Van Horn responded that integrity means being honest with yourself and with everyone else, and trying to get to the truth and then relying on the truth to guide your decisions and actions. Virgil Moore noted that you should "always associate yourself with the highest standards of scientific integrity that you can find, whether it be your employer, your professional society, or continuing education." Whatever the source or sources, it is critical that you have a strong moral compass and that you never deviate from your core values. Time, money, convenience, or agency politics may exert pressure, but it is imperative that you never compromise your integrity.

1. <u>Always practice the Golden Rule</u>—Respect all people and hurt no one in your quest for professional success. Jim Martin often quotes Harry Wagner, a former Oregon Department of Fish and Wildlife fisheries chief, who said, "The test of a professional is not what your colleagues think of you, but what your opponents think of you." Surely this is the ultimate test of living by the Golden Rule.

Larry went on to describe some ways that AFS could help fisheries professionals address each of these 10 keys to success. The American Fisheries Society facilitates becoming knowledgeable through its journals, meetings, newsletters, books, and special publications. Involvement in AFS provides numerous opportunities to demonstrate flexibility and receive feedback from some of the most noted authorities in the profession. Most of these "giants" in AFS will willingly share their time and abilities, allowing others to adopt elements of their style. The American Fisheries Society activities are always infused with some degree of fun, whether it be socials at meetings, serving with other professionals on committees or task forces, or a variety of other opportunities. There are certainly ample ways to develop a work ethic through AFS. By simply volunteering to contribute time and abilities, a member can become instantly busy with tasks that will require a lot of hard work and perseverance. Almost everything in AFS is accomplished through some sort of teaming effort, so participation in such efforts is a perfect way to practice skills of working as part of a team. Perhaps most importantly, building professional networks is what AFS is all about. Participation in AFS activities helps members build a vast and valuable network of fellow professionals, and the AFS membership directory is an indispensable tool in maintaining this network. The American Fisheries Society meetings are designed to provide opportunities for members to strike a balance between professional activities and personal activities. One example is the organization of a 5-km road race, known as the Spawning Run, at each AFS Annual Meeting. The American Fisheries Society recognizes the importance of ethics and has a strong code of ethics that can help guide decision making in difficult situations. The American Fisheries Society professional certification is also a valuable tool in ensuring the highest ethical standards within the profession. Finally, by developing a vast network through AFS, members will be provided with unlimited opportunities to practice the Golden Rule. Put simply, AFS is the perfect means to launch and facilitate your professional career.

Sources of Additional Information

The AFS Standard of Professional Conduct, the current AFS Strategic Plan, and the current President's Program of Work, which aims to move AFS toward attaining its goals as set out in the Strategic Plan, can be found in the *About AFS* section of the AFS web site (http://www.fisheries.org). Information on how to sign up for the AFS-wide email listserv (AFS-L) is also available in that section. Complete details of the AFS Professional Certification Program are available at the web site under the *Certification* section. The library of PowerPoint presentations maintained by the Fisheries Conservation Foundation is available through its web site (http://www.fisheries.org/foundation).

References

Ames, E. P. 2004. Atlantic cod stock structure in the Gulf of Maine. Fisheries 29(1): 10–28.

Leopold, A. 1949. A Sand County almanac. Oxford University Press, New York.

Olmsted, L. L. 2005a. How to get really smart and secure a super job. Fisheries 30(3): 26–27.

Olmsted, L. L. 2005b. How to be wildly successful. Fisheries 30(4): 27–28.